Praise for *Design for How People Think*

"This book takes John's years of research and practice and turns them into an accessible, practical, and enjoyable read with a playful sense of humor. Regardless of the role you play on your team, the Six Minds framework and discovery techniques will help you unlock key insights about your customers and propel your product's success."

HEATHER WINKLE,
MANAGING VICE PRESIDENT OF DESIGN AT CAPITAL ONE

"Design for How People Think *reads just like a conversation with John—clear, engaging, and always quick to the point. This is a terrific book for people new to UX research or who work with UX and want to better understand its role in the product design process. Even long-time practitioners will find the Six Minds framework new and useful and the summary of key concepts, a helpful refresh. Lots of great examples and concrete, practical advice.*"

LAURA CUOZZO GUARNOTTA,
USER EXPERIENCE RESEARCH LEAD AT GOOGLE

"The design industry is changing quickly, and with AI and ML, new tools will replace design asset delivery once crafted with a keyboard and mouse. We have the tools to synthesize research now, but Design for How People Think *helps you to understand how to synthesize for design of the future, focusing deeply on how customers think, not on how they use our products.*"

JASON WISHARD,
DIRECTOR, DESIGN PRACTICE MANAGEMENT, AT CAPITAL ONE,
CONSUMER BANK DESIGN

"The demand for world-class customer experiences is increasing daily. Unfortunately, with new technologies like AR and AI, the rules of customer experience have changed. John's book provides a way forward to understand how the brain processes these technologies, drawing a scientific road map of how to deliver customer experiences that work."

JASON PAPPAS,
INNOVATION AND DIGITAL TRANSFORMATION LEADER, AT EATON

Design for How People Think

Using Brain Science to Build Better Products

John Whalen, PhD

Beijing · Boston · Farnham · Sebastopol · Tokyo

DESIGN FOR HOW PEOPLE THINK
by John Whalen, PhD

Copyright © 2019 John Whalen. All rights reserved.

Published by O'Reilly Media, Inc., 1005 Gravenstein Highway North, Sebastopol, CA 95472.

O'Reilly books may be purchased for educational, business, or sales promotional use. Online editions are also available for most titles (*http://oreilly.com*). For more information, contact our corporate/institutional sales department: (800) 998-9938 or *corporate@oreilly.com*.

Development Editor: Angela Rufino
Acquisition Editor: Jessica Haberman
Production Editor: Katherine Tozer
Copyeditor: Jasmine Kwityn
Proofreader: Rachel Head
Indexer: Lucie Haskins

Cover Designer: Karen Montgomery
Interior Designers: Ron Bilodeau and Monica Kamsvaag
Illustrators: Rebecca Demarest

April 2019: First Edition

Revision History for the First Edition:

2021-09-03	Eighth Release
2022-12-02	Ninth Release

See *https://www.oreilly.com/catalog/errata.csp?isbn=0636920077916* for release details.

978-1-491-98545-8

[LSI]

[*contents*]

Why I Wrote This Book

"A Psychologist Doing Product and Service Design? How Interesting..."

A COMMON RESPONSE WHEN I introduce myself as a psychologist who does product design is surprise: "Isn't that the job of *designers*? Oh, you must really get into the customer's head! Are you analyzing me right now?" [No comment! ;)]

While often intrigued, these people don't know how knowledge about human cognition and emotion can be applied in digital product and service design. They are not alone. After giving a talk at SXSW, I had more than one person say, "That is so cool! I wish I knew how to use that in my products..."

So Do You Want the Secret to Designing a Great Experience?

Start by thinking of a truly great experience in your life. Was it one of life's milestones? The birth of a child, marriage, graduation, etc.? Or was it a specific moment in time—a concert with your favorite band, a play on Broadway, an immersive dance club, an amazing sunset by the ocean, or watching your favorite movie?

You might remark that it was "brilliant" or "an amazing experience" to a friend.

What you probably didn't think about was how many different senses and cognitive processes blended together to make that experience for you. Can you almost smell the popcorn when you think of that movie? Maybe the play had not only great acting but creative costumes and lighting and starred someone you thought was good-looking and moved with amazing grace. Was it the dancing with festive fans nearby? So many elements come together to provide a "singularly" great experience.

How might you go about designing a great experience for your product or service? What are the sensations, emotions, and cognitive processes that make up your experience? How can you tease them apart systematically into component parts? How will you know you are building the right thing?

This book is designed to help you understand and harness what we know about human psychology to unpack experiences into their component parts and uncover what is needed to build a great experience. This is a great time to do so. The pace of scientific discovery in brain science has been steadily increasing. There have been tremendous breakthroughs in psychology, neuroscience, behavioral economics, and human–computer interaction that provide new information about distinct brain functions and how humans process that information to generate that feeling of a single experience.

How Humans Think About Thinking (And What We Don't Realize)

Your thoughts about your own thinking can be misleading because there are limits to your awareness of your own mental processes. We all know what it's like to struggle over a decision about which outfit to wear for a big date or a job interview: Will you meet their initial expectations? Will they get the wrong impression? Does it look good? Do you look professional enough? Are those shoes too attention-grabbing? There are a lot of thoughts there—but there are still more thoughts that you are unable to articulate, or are even aware of.

One of the fascinating things about consciousness is how much of our thinking is impenetrable to our own awareness. For example, while we are easily able to identify the shoes we plan to wear to an interview, we do not have insight into *how* we recognized the shoes as shoes, or how we were able to sense the color of the shoes. We generally don't know where our eyes are moving to next, the position of our tongue (yikes!), how we control our heart rate, how we see, how we recognize words, or how we remember our first home (or anything), to mention just a few examples. As a result, we must identify and understand not only consciously accessible cognitive processes, but also those that are unconscious (like eye movements often are) or deep-seated—like the emotions related to those concepts.

I was trained in my PhD program as a cognitive scientist, studying memory, language, problem solving, and decision making. Now, after more than 15 years of consulting, I've learned how to interview and observe customers, learn what makes them tick on the inside, and identify opportunities to make exceptional products or services that grow businesses and provide a great experience for customers. I now work with some of the world's biggest companies influencing product strategies for global products. I hope you benefit from what I'm sharing here and enjoy the process of understanding your customers as much as I do!

Who This Book Is For

I wrote this book to help product owners, product managers, designers, user experience professionals, and developers to: (a) identify the cognitive processes that together form a brilliant experience, (b) learn how to extract information about these through contextual interviews with your customers, and (c) apply that knowledge in your product and service design processes. This is meant to be a practical and hands-on book, not an academic one.

Why Product Managers, Designers, and Strategists Need This Information

No product, service, or experience will ever be a runaway success if it does not end up meeting the needs of the target audience. You want someone exposed to your product or service for the first time to say something like a Londoner might: "Right, that's brilliant!"

But how, as a corporate leader, marketer, product owner, or designer, can you be sure that your products or services will create an exceptional experience? You can ask customers what they want, but many don't know what they need or can't clearly articulate their needs. You might work from the vantage point of what *you* would want, but do you really know how a 13-year-old girl wants to work with her "Insta" and "Finsta" (Instagram)? How a high-net-worth investor wants to "seek alpha"? Or how a 75-year-old attorney wants to search for tax law regarding reverse triangular mergers? So how should you proceed?

This book is designed to equip you with the tools you need to deeply understand your customers' needs and perspective. As a cognitive scientist, I feel like "usability testing" and "market surveys" and "empathy research" are at times both too simplistic and too complicated. I think they sometimes miss the mark in helping you—the product team—to understand what you need to build.

I believe there is a better way: by understanding the elements of an experience (in this book I will describe six as a start), you can better identify audience needs at different levels of explanation. Throughout this book, I'll help you better understand what the audience needs at those different levels and make sure you hit the mark with each one.

How This Book Is Organized

PART I: RETHINKING "THE" EXPERIENCE

Part I is designed to share some of the fascinating properties of human cognition that you as designers, product managers, and developers need to be aware of:

- Chapter 1 introduces the notion that "an experience" is actually many different experiences and cognitive processes all rolled into one human experience.

- Chapter 2 gets you thinking about vision and attention—what draws you in, what you are seeking, and how much of your thinking happens without your conscious awareness.

- Chapter 3 reminds you that a huge part of your brain is wired to help you represent space, and gets you thinking about how you might harness that machinery in your virtual space (e.g., an app or website). Did I mention the part about Tunisian ants in the desert? Have a look!

- Chapter 4 is there to emphasize how much of your experience is actually manufactured and filled in by your memories, and how quickly you go from concrete objects to abstract thoughts. What are your customers filling in with their thoughts?

- Chapter 5 reminds you that you aren't your customer. Your customers rarely use the language you do, and you can quickly lose their trust by being either too simplistic or overly technical with the words you use. And do the words you use mean the same thing that your customers think they mean?

- Chapter 6 gets to what we typically think of when we are thinking: solving problems and making decisions. However, this serves as a reminder that in many cases (escape rooms are a good example) what we *think* we're trying to solve is often not what we actually have to solve. What problem do your customers *think* they need to solve with your product or service?

- Chapter 7 describes how our best intentions for wise decisions in Chapter 6 are often co-opted by our emotional selves. What will appeal to your customers, enhance their lives, and awaken their deepest passions—and allay their deepest fears?

After reading Part I, you will (hopefully) know much more about human cognition and how an experience is composed of many thoughts, cognitive processes, and emotions than you did before.

PART II: EXPOSING SECRETS

Part II is designed to make every member of your team a valuable member of the customer research team. This part shows you how you can watch your customers work, and interview them, and in doing so expose valuable insights about the cognitive processes described in Part I. This is practical, "boots-on-the-ground" stuff. You do not need to be a psychologist to do this!

- Chapter 8 introduces how I want you to conduct what I am calling a *contextual interview*—a hybrid of a simple interview and watching someone work (what researchers often call *contextual inquiry*). This chapter covers a lot, including: Why do interviews at all? What do I need to capture? And once I have all my notes, how do I organize them to get product insights out of them?

- Chapter 9 helps teach you how to gather lots of valuable insights about what captured your customers' attention, what they were seeking, and why. I'll share how I used this very same technique to help security teams at major buildings and stadiums keep people safer by better managing all the cameras and bells and whistles and beeps constantly alerting them to everything from open doors to stuck elevators to faulty water heaters!

- Chapter 10 shows you how to carefully record the words your customers are using, and what they mean to them. You'll learn how we helped to organize every single malady at *NIH.gov* for both world experts and ordinary folks—a common challenge at many organizations.

- Chapter 11 gets you thinking about your customers' mental model for your product or service. Where do they think they are in your app or service? What do *they* think they should do to move from step to step?

- Chapter 12 reminds you to harness what your customers already know. What knowledge are they bringing with them? How do they think your product or service works? What experiences inform that? I share the example of designing products and services for small business owners and quickly realizing there are two hugely different groups with completely different needs, suggesting two different sets of products and services should be offered.

- Chapter 13 helps you to discover what your customers think they are trying to solve, and what they think they can do about it. You'll see that part of a great experience might be helping your customers to realize that they actually have a very different problem to solve. I describe why first-time home buyers are often a great example fo this.

- Chapter 14 helps you intuit what was never spoken in the interview. What are your customers' biggest goals? What are their fears? What do they need to know to be able to say yes to your product or service? I'll describe how interviews asking first about what credit cards are in a customer's wallet can quickly turn into revelatory experiences for them (hugs may be coming!). This is when you might realize that you really need to refocus your product line to help your customers achieve their biggest life goals.

PART III: PUTTING THE SIX MINDS TO WORK IN YOUR DESIGNS

OK, you've discovered some fascinating insights into what attracts your customers, the words they use, the emotions they have, the problems they are trying to solve, and more. But how does this change your product? Read on!

- Chapter 15 is all about "sense-making": how to identify patterns in your data and how you might segment customers by using what you know about their thought patterns and emotions—which might be a very different way of thinking about your customers than focusing on their zip codes, average sales, or years of experience! You'll find out how we put this to work for groups as diverse as Millennials managing their money and families that have experienced fraud.

- Chapter 16 gets you thinking about how to make your product a success for each of the groups you identified in Chapter 15 by marketing it appropriately. How do you *appeal* to them by recognizing what they think they need, *enhance* their lives, and ultimately *awaken* their passions and helping them to achieve their biggest goals in life?

- Chapter 17 covers how to test your product or service idea. Why not get to success and launch faster? Learn how we integrate the Six Minds into a lean, Agile approach (my apologies to any of the buzz words I may fail to mention).

- Chapter 18 is a summary of sorts. I want to show you how my complany launched some of the top one hundred websites in the world and had the Six Minds in mind as we designed. I also want you to think about how the Six Minds aren't static. The elements that are most crucial may change over time (e.g., during the buying process).

- Chapter 19 is forward-thinking. You can't swing a cat in Silicon Valley lately without having an artificial intelligence (AI) or machine learning (ML) strategy (no cats were harmed in the writing of this book). I encourage all of you, especially product owners and technical leads, to take a step back and think about what you are really trying to accomplish. I argue that knowing more about the humans you will be interacting with will increase the likelihood that your costly and risky endeavor is a smashing success.

Think about how ML and AI can support humans so they attend to the right information, get fed the right words at the right time, and ultimately make better decisions and solve more problems.

Now go! Read on, and with your new knowledge, tools, and skills make the best products and services your customers will ever experience!

Conventions Used in This Book

The following typographical conventions are used in this book:

Italic indicates new terms, URLs, email addresses, filenames, and file extensions.

[SIDE NOTE]

This element signifies a note or tip.

Warning

This element signifies a warning.

O'Reilly Online Learning

For almost 40 years, O'Reilly Media has provided technology and business training, knowledge, and insight to help companies succeed.

Our unique network of experts and innovators share their knowledge and expertise through books, articles, conferences, and our online learning platform. O'Reilly's online learning platform gives you on-demand access to live training courses, in-depth learning paths, interactive coding environments, and a vast collection of text and video from O'Reilly and 200+ other publishers. For more information, please visit *http://oreilly.com*.

How to Contact Us

Please address comments and questions concerning this book to the publisher:

O'Reilly Media, Inc.
1005 Gravenstein Highway North
Sebastopol, CA 95472
800-998-9938 (in the United States or Canada)
707-829-0515 (international or local)
707-829-0104 (fax)

We have a web page for this book, where we list errata, examples, and any additional information. You can access this page at *http://bit.ly/design-how-people-think*.

To comment or ask technical questions about this book, send email to *bookquestions@oreilly.com*.

For more information about our books, courses, conferences, and news, see our website at *http://www.oreilly.com*

Find us on Facebook: *http://facebook.com/oreilly*

Follow us on Twitter: *http://twitter.com/oreillymedia*

Watch us on YouTube: *http://www.youtube.com/oreillymedia*

Acknowledgments

I owe a debt of gratitude to my colleagues at Brilliant Experience, especially those who got me to start writing *and* to finish. To my friends and colleagues at User Experience Professionals Association, nationally and here in Washington, DC, you inspire me every single day. I hope you find this helpful! To my editors and the team at O'Reilly: you have been patient and helpful whether I deserved it or not. Thank you! And to my family, who might have wondered what I was doing as I typed away in the office or coffee shops all those hours, I'm back!

A Final Note to the Psychologists and Cognitive Scientists Reading This

Bear with me. In a practical and applied book I simply can't get to all the nuances of the mind/brain that exist, and I need a way to communicate to a broad audience what is relevant to product and service design. There are a myriad of amazing facts about our minds which (sadly) I am forced to gloss over, but I do so intentionally so that we may focus on the broader notion of designing with multiple cognitive processes in mind, and ultimately allow for an evidence-based and psychologically driven design process. It would be an honor to have my fellow scientists work with me to integrate more of what we know about our minds into the design of products and services. I welcome your refinements. At the end of each chapter I will point to further citations the interested reader can pursue to get more of the science they should know.

Let Me Know What You Think

The conversation is just beginning. Google me! Share your thoughts and help me refine my thinking.

Part I

Rethinking "the" Experience

IT'S PRETTY COMMON FOR people to ask, "What do you do?" When I tell someone I'm a psychologist, they *think* they know what I do, but when I tell them I'm a cognitive scientist they know they *don't* know what I do.

Generally, cognitive science is the study of cognition–thinking–and all the mental processes that go into recognizing objects, using a language, reasoning, and problem solving. I believe you will discover a new and valuable reframing of what an experience is (and how to design one).

While we all experience consciousness, there are a myriad of cognitive processes that are highly automated and subconscious. For example, how is it that you just know a chair is a chair? Your visual system identifies figure from ground, composes three dimensions from a two-dimensional image on the back of your eye, and eventually relates that image to other ones you've stored in memory and relates that concept to a linguistic element ("chair").

If there are that many steps in recognizing a chair–each with their own specialized processing systems–we should consider the processes that compose an experience. In Part I of this book, I propose while an event might be "an experience" to us consciously, it is actually a symphony of many different cognitive processes in the brain.

By looking at each one in turn, we can identify the components of "an experience," and what we need to build to generate a new one. There are almost certainly hundreds of distinct processes, but in the next few chapters I distill this down to six cognitive processes that are the most relevant to product and service design: Vision/Attention, Wayfinding, Memory, Language, Decision Making, and Emotion.

Let's take a quick trip into your customer's conscious and subconscious thoughts now!

[1]

The Six Minds of Experience

SURELY IT IS THE case that there are hundreds of cognitive processes happening every second in your brain. But to simplify to a level that might be more relevant to product and service design, I propose that we limit ourselves to a subset that we can realistically measure and influence.

What are these processes and what are their functions? Let's use a concrete example to explain them. Consider the act of purchasing a chair suitable for your mid-century modern house. Perhaps you might be interested in a classic design from that period, like the Eames chair and ottoman shown in Figure 1-1. You are seeking to buy this online and browsing an ecommerce site.

FIGURE 1-1
Eames chair and ottoman

Vision, Attention, and Automaticity

As you first land on the furniture website to look for chairs, your attention and eyes might be drawn to the pictures to make sure you are on the right site. You might choose to look for the search option to type in "Eames chair." You might also scan the page for words such as "furniture," or for the word "chair," from which you might look for the appropriate category of chair. If you don't find "chair," you might look for other words that might represent a category that includes chairs. Let's suppose on scanning the options shown in Figure 1-2, you pick the "Living" option.

New Living Dining Bedroom Workspace Outdoor Storage Lighting Rugs Accessories Designers Sale

FIGURE 1-2
Navigation from Design Within Reach website

Wayfinding

Once you believe you've found a way into the site, the next task is to understand how to navigate its virtual space. While in the physical world we've learned the geography in and around our homes and how to get to some of our most frequented locations like our favorite grocery store or coffee shop, the virtual world may not always present our minds with the navigational cues that our brains are designed for— notably three-dimensional space.

Often it is the case that we aren't sure where we are within a website, app, or virtual experience. Nor do we always know how to navigate a virtual space. On a web page you might think to click on a term, like "Living" in the example in Figure 1-2. But in other cases, like Snapchat or Instagram, many people over the age of 18 might struggle to understand how to get around by swiping, clicking, or even waving their phones. Understanding where you are in space (real or virtual) and how you can navigate through that space (moving in 3D, swiping or tapping on phones) are critical to a great experience.

Language

I find when I'm around interior designers, I start to wonder if they speak a different language than I do. The words defining a category can vary dramatically based on your level of expertise. If you are an interior design expert, you might masterfully navigate a furniture site because you know what distinguishes an egg chair, swan chair, diamond chair, and lounge chair. In contrast, if you are new to the world of interior design, you might need to Google all these names to figure out what they are talking about! To create a great experience, we must understand the words our audience uses and meet them at the appropriate level. Asking experts to simply look up the category "chair" (far too imprecise) is about as helpful as asking someone who isn't a neuroscientist about the differences between the dorsolateral prefrontal cortex and anterior cingulate gyrus (both of which are neuroanatomical areas).

Memory

As I navigate an ecommerce site, I also have expectations about how it works. For example, I might expect that such a site will have a search box (and search results), product category pages (chairs), product pages (a specific chair), and a checkout process. We have expectations for any number of concepts. We automatically build mental expectations about people, places, processes, and more. As product designers, we need to make sure we understand what our customers' expectations are, and anticipate confusions that might arise if we deviate from those norms (e.g., think about how strange it felt the first time you left a Lyft or Uber or limousine without physically paying the driver).

Decision Making

Ultimately, you are seeking to accomplish your goals and make decisions. In this case you might be deciding if you should buy this chair (Figure 1-3). There are any number of questions that might go through your head as you make that decision. Would this look good in my living room? Can I afford it? Would it fit through the front door? At nearly $5,000, what happens if it is scratched or damaged during transit? Am I getting the best price? How should I maintain it? As product and service managers and designers, we need to think about all the steps along an individual customer's mental journey and be ready to answer the questions that come up along the way.

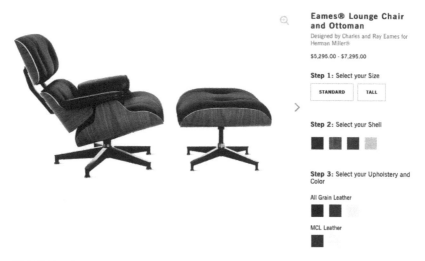

FIGURE 1-3
Product detail page from Design Within Reach

Emotion

While we may like to think we can be like Spock from *Star Trek* and make decisions completely logically, it has been well documented that a myriad of emotions affect both our experience and our thinking. Perhaps as you look at this chair you are thinking about how your friends would be impressed, or how it might show your status and that you've "made it." Or perhaps you're thinking "How pretentious!" or "$5,000 for a chair—how am I going to pay for that, rent, and food?!" and starting to panic. Identifying underlying emotions and deep-seated beliefs is critical to building a successful experience.

The Six Minds

Together, the very different processes described in preceding sections (shown in Figure 1-4), which are generally located in unique brain areas, come together to create what each of us perceive as a singular experience. While my fellow cognitive neuropsychologists would quickly agree that this is an oversimplification of both human anatomy and processes, there are some reasonable overarching themes that make this a level at which we can connect product design and neuroscience.

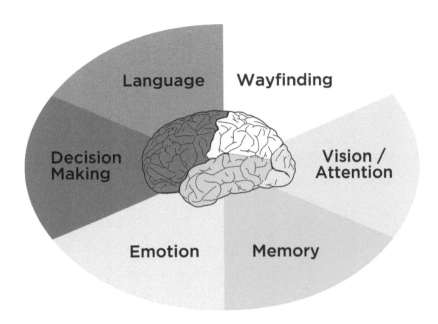

FIGURE 1-4
Six Minds of Experience

I think we all might agree that "an experience" is not singular at all, but rather is multidimensional, nuanced, and composed of many brain processes and representations. *The customer experience doesn't happen on a screen, it happens in the mind.*

Activity

Let me recommend you take a brief pause in your reading and go to an ecommerce website—ideally, one that you've rarely used—and search for books on the topic of "customer experience." When you do, do so in a new and self-aware state:

Vision/Attention
> Where did your eyes travel to first on the site? What were you looking for (e.g., images, colors, words)?

Wayfinding
> Did you know where you were on the site and how to navigate it? Were you ever uncertain? Why?

Language

> What words were you looking for? Did you experience terms you didn't understand, or were the categories ever too general?

Memory

> How were your expectations about how the site would work confirmed or violated?

Decision Making

> What were the microdecisions you made along the way as you sought to accomplish your goal of purchasing a book?

Emotion

> What concerns did you have? What might stop you from making a purchase (e.g., security, trust)?

Now that you have some sense of the mental representations you need to be aware of, you may ask: How do I, as a product manager, not a psychologist, determine where someone is looking and what they are looking for? How do I know what my product audience's expectations are? How can I expose deep-seated emotions? We'll get there in Part II of the book, but for right now I want to agree on what we mean by vision/attention, wayfinding, memory, language, emotion, and decision making. I want you to know more about each of these so you can recognize these processes "in the wild" as you observe and interview your customers.

[2]

In the Blink of an Eye: Vision, Attention, and Automaticity

From Representations to Experiences

THINK OF A TIME when you were asked to close your eyes for a big surprise (no peeking!), and then opened your eyes for the big reveal. At the moment you opened your eyes, you were taking in all kinds of sensations: light and dark areas in your scene, colors, objects (cake and candles?), faces (family and friends), sounds, smells, emotions (joy?). It is a great example of how instantaneous, multidimensional, and complex an experience can be.

Despite the vast ocean of input streaming in from our senses, we have the gift of nearly instant perception of an enormous portion of any given scene. It comes to us so naturally, yet is so difficult for a machine or a self-driving car. Upon reflection, it is amazing how "effortless" these processes are. They just work. You don't have to think about how to recognize objects or make sense of the physical world in three dimensions, except in very rare circumstances (e.g., dense fog).

These automatic processes start with neurons in the back of your eyeballs, with input passing through your corpus callosum to the back of your brain in the occipital cortex, then your temporal and parietal lobes in near real time. In this chapter we'll focus on the "what" and in the next we'll focus on the "where" (Figure 2-1).

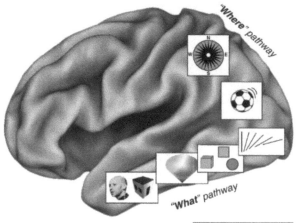

TRENDS in Cognitive Sciences

FIGURE 2-1
What/where pathways

With almost no conscious control, your brain brings together separate representations of brightness, edges, lines, line orientation,
color, motion, objects, and space (in addition to sounds, feelings, and
proprioception) into a singular experience. You have no conscious
awareness of these as separate and distinct representations, or that they
were brought together into a single experience, or that past memories
influence your perception, or that they evoke certain emotions.

This is a nontrivial accomplishment. It is incredibly difficult to build
a machine that can correctly make even the most basic distinctions
between objects that have similar colors and shapes—for example,
between muffins and Chihuahuas—which with a brief inspection you,
as a human, will get correct every time (Figure 2-2).

There are many, many things I could share about vision, object recognition, and perception, but the most important for our digital product
design purposes are these: (a) there are many processes taking place
simultaneously of which you have little conscious awareness or control,
and (b) many computationally challenging processes are taking place
constantly that don't require conscious mental effort on your part.

FIGURE 2-2
Muffins or Chihuahuas?

In Nobel Prize winner Daniel Kahneman's fantastic book *Thinking, Fast and Slow,* he makes the compelling point that there are two very different ways in which your brain works. You are aware of and in conscious control over the first set of processes ("thinking slow"). And you have little to no conscious control or introspection over the other set of automatic processes ("thinking fast").

When designing products and services, we as designers are often very good at focusing on the conscious processes (e.g., decision making), but we rarely design with the intention of harnessing our fast automatic processes. They occur quickly and automatically, and we essentially "get them for free" in terms of the mental effort our audience needs to expend as it uses them. As product designers, we should harness both these conscious and automatic processes because they are relatively independent. The latter don't meaningfully tax the former. In later chapters, we'll see exactly how to do so in detail, but for now let's discuss one good example of an automatic visual process we can harness: visual attention.

Unconscious Behaviors: Caught You Looking

Think back to the vignette I gave you at the start of the chapter: opening your eyes for that big surprise. If you try covering your eyes now and suddenly uncovering them, you may find that your eyes dart around the scene. In fact, that is consistent with your standard eye movements. Eyes don't typically move in a smooth pattern. Rather, they jump from location to location (a phenomenon we call *saccades*). This motion can be measured using specialized tools like an infrared eye tracking system, which can now be built into specialized glasses (Figure 2-3) or a small strip under a computer monitor (Figure 2-4).

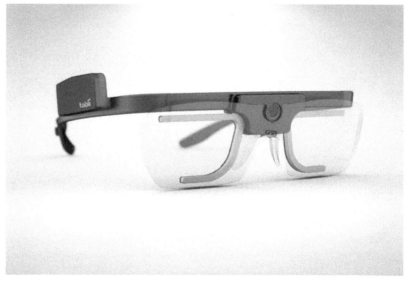

FIGURE 2-3
Tobii Glasses 2

FIGURE 2-4
Tobii X2-30 (positioned below the computer screen)

These tools have documented what is now a well-established pattern of eye movements on things like web pages and search results. Imagine that you just typed in a Google search and are viewing the results on a laptop. On average we tend to look 7 to 10 words into the first line of the results, 5 to 7 words into the next line, and even fewer words into the third line of results. There is a characteristic "F-shaped" pattern that our eye movements (saccades) form. Looking at the image in Figure 2-5, the more red the value, the more time was spent on that part of the screen.

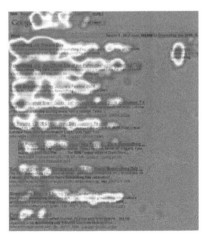

FIGURE 2-5
Heat map of search eye tracking "F" pattern (source: *http://bit. ly/2n6yQuw*)

Visual Popout

While humans are capable of controlling our eye movements, much of the time we let our automatic processes take charge. Having our eye movements on "autopilot" works well in part, because things in our visual field strongly attract our attention when they stand out from the other features in the visual scene. These outliers automatically "pop out" to draw our attention and eye movements.

As product designers, we often fail to harness this powerful automatic process, yet it's a great way to draw attention to the right elements in a display—think of the children's song from *Sesame Street* ("One of these things is not like the others, one of these things just doesn't belong..."). Some of the ways you can create visual pop out in a scene are demonstrated in Figure 2-6. Other important features I would add to this list are visual contrast (relative brightness and darkness) and motion. The items in the bottom-right corner "pop out" because they have both a unique attribute (e.g., shape, size, orientation) and a unique visual contrast relative to the others in their groupings.

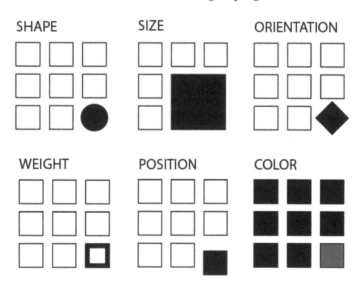

FIGURE 2-6
Visual popout

One interesting thing about visual popout is that the unique object draws attention regardless of the number of competing elements. In a complex scene (e.g., modern car dashboards), this can be an extremely helpful method of directing one's attention when needed.

If you are an astute reader thinking about the locus of control with eye movements, you might be wondering "*Who* decides where to look next if *you* aren't consciously directing your eyes?" and "How exactly do your eyes get drawn to one part of a visual scene?" It turns out that your visual attention system—the system used to decide where to take your eyes next—forms a blurry (somewhat Gaussian, for you Photoshop users) and largely black-and-white representation of the scene to make the decision. It uses that representation, which is constantly being updated, to decide where the locus of your attention should be, provided "you" are not consciously directing your eyes. To ponder: "Is it *you* when it is done automatically?"

As a designer you can anticipate where someone's eyes might go in a visual scene if you take that scene and use a program like Photoshop to turn down the color and squint your eyes (and/or use more than one Gaussian blur). That test will give you a pretty good idea of where people's eyes would be drawn in the scene, were you to measure their actual eye gaze pattern using an eye tracking device.

Oops, You Missed That!

One of the most interesting results you get from studying eye movements is the null result: what people *never* look at. For example, I've seen a web form design in which the designers tried to put helpful supplemental information in a column off to the side on the right of the screen—exactly where ads are typically placed. Unfortunately, we as consumers have been trained to assume that information on the right side of the screen is an ad and as a result will simply ignore anything in that location (helpful or not). Knowing about past experiences will surely help us to anticipate where people are looking and help us to craft designs in a way that actually directs—not repels—attention to the helpful information.

If your customers never look at a part of your product or screen, then they will never know what is there. You might as well not have put the information there to begin with. However, when attentional systems are harnessed correctly through psychology-driven design, there is amazing potential to draw people's attention to precisely what they need. This is the opportunity we as product designers should always exploit to optimize the experience.

Our Visual System Creates Clarity
When There Is None

I can't resist sharing one more characteristic of human vision—specifically, about visual acuity. When looking at a scene, our subjective experience is that all parts of that scene are equally clear, in focus, and detailed. In actuality, both your visual acuity *and* your ability to perceive color drop off precipitously from your focal point (what you are staring at). Only about 2° of visual angle (the equivalent of your two thumbs at arm's-length distance) are packed with neurons and can provide both excellent acuity and strong color accuracy.

Don't believe me? Go to your closest bookshelf. Stare at one particular book cover and try to read the name of the book that is two books over. You may be shocked to realize you are unable to do so. Go ahead, I'll wait!

Just a few degrees of visual angle from where our eyes are staring (foveating), our brains make all kinds of assumptions as to what is there, and we are unable to fully process it. This makes where you are looking turn out to be crucial for an experience. Nearby just doesn't cut it!

Ceci N'est Pas une Pipe: Learning What
Someone Understands Something to Be,
Not What Might Actually Be There

Whether we present words on a page, image, or chart, the displayed elements are only useful to the extent the end users accurately identify the objects that they are seeing.

Icons are a particularly good example. If you ask someone who has never used Instagram what each of the icons in Figure 2-7 represent, I'm willing to bet they won't correctly guess what each icon means. For that person that icon *is* the meaning they put on it at that moment (regardless of what it was *meant* to represent). As a design team, it is essential to test all of your visual elements and make sure that they are widely identified correctly or, if absolutely needed, that they can be learned with practice. When in doubt, do not battle standards to be creative. Go with the standard icon and be unique in other ways.

FIGURE 2-7
Instagram controls

Further Reading

Evans, J. S. B. T. (2008). "Dual-Processing Accounts of Reasoning, Judgment, and Social Cognition." *Annual Review of Psychology* 59: 255–278.

Evans, J. S. B. T., & Stanovich, K. E. (2013). "Dual-Process Theories of Higher Cognition: Advancing the Debate." *Perspectives on Psychological Science* 8(3): 223–241.

Kahneman, D. (2011). *Thinking, Fast and Slow*. New York: Macmillan.

[3]

Wayfinding: Where Am I?

A LOGICAL EXTENSION TO thinking about what we are looking at is understanding where we are in space. A large portion of the human brain is devoted to the representation of spatial information, so we ought to discuss it and consider how this cognitive process might be harnessed in our designs from two perspectives: knowing where we are, and knowing how we can move around in space.

The Ant in the Desert: Computing Euclidean Space

To help you think about the concept of wayfinding, I'm going to tell you about large Tunisian ants in the desert—who interestingly share an important ability that we have. I first read about this and other amazing animal abilities in Randy Gallistel's *The Organization of Learning*, which suggests that living creatures great and small share more cognitive processes than you might have anticipated. Representations of time, space, distance, light and sound intensity, and proportion of food over a geographic area are just a few examples of computations many creatures are capable of.

Imagine yourself as a Tunisian ant. Determining your location in a desert is a particularly thorny problem. There are no landmarks like trees, and the landscape can frequently change as sand moves in the wind. Therefore, ants that leave their nest must use something other than landmarks to find their way home. Their footprints, landmarks, and scent trails in the sand are all unreliable as they can change with a strong breeze.

Furthermore, these ants take meandering walks in the Tunisian desert scouting for food (the ant in Figure 3-1 generally goes northwest from its nest). In this experiment, a scientist has left out a bird feeder full of sweet syrup. This lucky ant climbs into the feeder, finds the syrup, and realizes that it has just found the jackpot of all food sources. After sampling the syrup, it can't wait to "tell" its fellow ants about the great news! However, before it can do this, the scientist picks up the feeder (with the ant inside) and moves it east about 12 meters (depicted by the red arrow in the diagram).

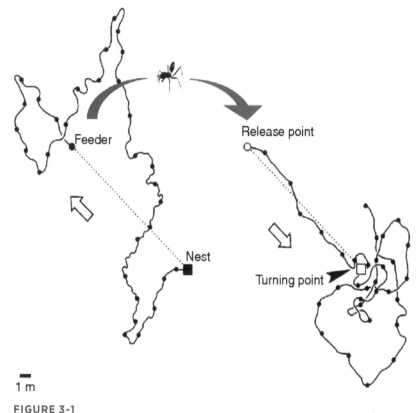

FIGURE 3-1
Tunisian ant in the desert

The ant, still eager to share the good news with everyone at home, attempts to make a beeline (or "antline") back home. The ant heads straight southeast, almost exactly in the direction where the anthill should have been, had the feeder not been moved. It travels approximately the distance needed, then starts walking in circles to spot the nest (which is a sensible strategy given there are no landmarks). Sadly, this particular ant doesn't take into consideration having been picked up, and so is off by exactly the amount that the experimenter moved the feeder.

Nevertheless, this pattern of behavior demonstrates that the ant is capable of computing the net direction and distance traveled in Euclidean space (using the sun, no less). This is a great example of what our parietal lobes are great at computing.

Locating Yourself in Physical and Virtual Space

Just like that ant, we all have to determine where we are in space, where we want to go, and what we must do in order to get to our destination. We do this using our brain's "where" system—one of the largest regions in the mammalian cortex.

If we have this uncanny, impressive ability to map space in the physical world built into us, wouldn't it make sense if as product and service designers we tapped into its potential when it comes to wayfinding in the digital world?

[SIDE NOTE]

If you characterize yourself as "not good with directions," you might be pleasantly surprised to find you're better than you realize. Consider, for example, how effortlessly you walk to the bathroom in the morning from your bed without thinking about it. And, if it is of any solace, know that like the ant, we were never designed to be picked up by a car and transported into the middle of a parking lot that has very few unique visual cues from which to find that car on our return trip home.

As I talk about "wayfinding" in this book, keep in mind that I'm linking two concepts which are similar, but do not necessarily harness the same underlying cognitive processes:

- Human wayfinding skills in the physical world using 3D space and motion over time

- Wayfinding and travel in the virtual world

There is an overlap between the two, but as we study this more carefully, we'll see that this is not a simple one-to-one mapping. The virtual world in most of today's interfaces on phones and web browsers strips away many wayfinding landmarks and cues. It isn't always clear where we are within a web page, app, or spoken experience (Alexa, Siri, etc.), nor is it always clear how to get where we want to be (or even create a mental map of where we are). Yet understanding where you are and how to move around the environment (real or virtual) is clearly critical to a great experience.

Where Can I Go? How Will I Get There?

In the physical world, it's hard to get anywhere without distinct cues. Gate numbers at airports, signs on the highway, and trail markers on a hike are just a few of the tangible "breadcrumbs" that (most of the time) make our lives easier.

Navigating a new digital interface can be like walking around a shopping mall without a map: it is easy to get lost because there are so few distinct cues to indicate where you are in space. Figure 3-2 shows a picture of a mall near my house. There are about eight hallways that are nearly identical to this one. Just imagine your friend saying "I'm near the tables and chairs that are under the chandeliers" and then trying to find them!

FIGURE 3-2
Westfield Montgomery mall

To make things even harder, unlike in the real world, where we know how to locomote by walking, in the digital world, the actions we need to take to get to where we are going sometimes differ dramatically between products (e.g., apps versus operating systems). You may need to tap your phone for the desired action to occur, shake the whole phone, hit the center button, double tap, control-click, swipe right, etc.

Some interfaces make wayfinding much harder than it needs to be. Many (older?) people find it incredibly difficult to navigate around Snapchat, for example. Perhaps you are one of them! In many cases, there is no button or link to get you from one place to another, so you just have to know where to click or swipe to get places. It is full of hidden "Easter eggs" that most people (Gen Y and Z excepted) don't know how to find (Figure 3-3).

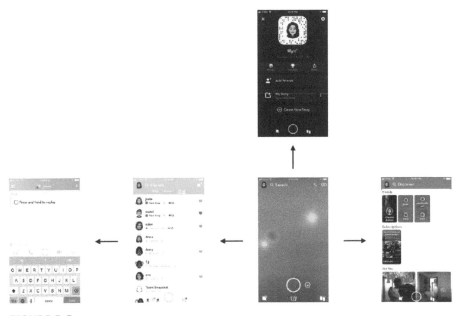

FIGURE 3-3
Snapchat navigation

When Snapchat was updated in 2017, there was a mass revolt from the teens who loved it (don't believe me? Google it!). Why? Because the users' existing wayfinding expectations no longer applied. As I write this book, Snapchat is working hard to unwind those changes to conform better to existing expectations. Take note of that lesson as you design and redesign your products and services: matched expectations can make for a great experience, and violated expectations can often destroy an experience.

The more we can connect our virtual world to some equivalency of the physical world, the better our virtual world will be. We're starting to get there, with augmented reality (AR) and virtual reality (VR), or even cues like edges of tiles that protrude from the edge of an interface (like Pinterest's images, which protrude from the edge of the screen) to suggest a horizontally scrollable area. But there are so many more opportunities to improve today's interfaces! Even something as basic as virtual breadcrumbs or cues (e.g., a slightly different background color for each section of a news site) could serve us well as navigational hints (that goes for you too, Westfield Montgomery mall!).

One of the navigational cues we cognitive scientists believe product designers underuse is our sense of 3D space. While you may never need to "walk" through a virtual space, there may be interesting ways to use 3D spatial cues, like in the scene shown in Figure 3-4. This scene provides perspective through the change in size of the cars and the width of the sidewalk as it extends back. This is an automatic cognitive processing system that we (as designers and humans) essentially "get for free." Everyone has it. Further, this part of the "fast" system works automatically without taxing conscious mental processes. A myriad of interesting and as yet untapped possibilities abound!

FIGURE 3-4
Visual perspective

Testing Interfaces to Reveal Metaphors for Interaction

One thing that we do know today is that it is crucial to test interfaces to see if the metaphors we have created (for where customers are and how customers interact with a product) are clear. One of the early studies done using touchscreen laptops demonstrated the value of testing to learn how users think they can move around in the virtual space of an app or site. When attempting to use these devices for the first time, users instinctively used metaphors from the physical world, as you can see in Figure 3-5. Participants touched what they wanted to select (upper-right frame), dragged a web page up or down like it was a physical scroll (lower left frame), and touched the screen in the location where they wanted to type something (upper-left frame).

FIGURE 3-5
First reactions to a touchscreen laptop

However, in addition to simply doing what might be expected, as in every product test I've ever conducted, the test also uncovered things that we did not expect (Figure 3-6).

In this example, the participant used both thumbs on the monitor while resting his hands on the sides of the monitor, sliding the interface up and down using both thumbs on either side of the screen. Who might have predicted that?!

FIGURE 3-6
Using a touchscreen laptop with two thumbs

The touchscreen test demonstrated two things:

- We can never fully anticipate how customers will interact with a new tool, which is why it's so important to test products with actual customers and observe their behavior.

- It's crucial to learn how people represent virtual space, and which interactions they believe will allow them to move around in that space.

While doing so, you are observing those parietal lobes at work!

While observing users interact with relatively "flat" (i.e., lacking 3D cues) on-screen television apps like Netflix or Amazon Fire, we've also learned not only about how they try to navigate the virtual menu options, but also what their expectations are for that space.

In the real world, there is no delay when you move something. Naturally, then, when users select something in virtual space, they expect the system to respond instantaneously. If (as in the case shown in Figure 3-7) nothing happens a few seconds after you (virtually) "click" on an object, your brain is naturally puzzled, and as a result you instinctively focus on that oddity, removing you from the intended virtual experience.

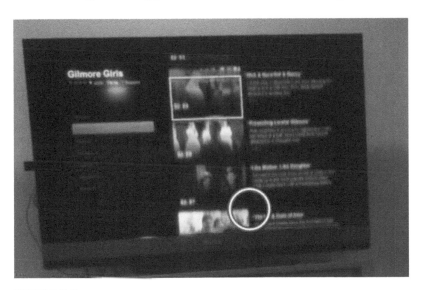

FIGURE 3-7
Eye tracking TV screen interface

Thinking to the Future: Is There a "Where" in a Voice Interface?

There is great potential for voice-activated interfaces like Google Home, Amazon Alexa, Apple Siri, Microsoft Cortana, and more. But in our testing of these voice interfaces, we've found new users often demonstrate anxiety around these devices because they lack any physical cues that the device is listening to and/or hearing them, and because the system's interactions and timing are so different from what customers expect of humans.

In testing both business and personal uses for these tools in a series of head-to-head comparisons, we've found there are a few major challenges that lie ahead for voice interfaces. First, unlike in the real world or with screen-based interfaces, there are no cues about where you are in the system. Suppose you start to discuss the weather in Paris with a voice interface. You might ask a follow-up question like "How long does it take to get to Monaco?" You're still thinking about Paris, but it's not clear if the voice system's frame of reference is still Paris. Today, with only a few exceptions, these systems start fresh in every discussion and rarely follow a conversational thread (e.g., that you are still talking about Paris when you ask about getting to Monaco).

Second, if the system jumps to a specific topical or app "area" (e.g., Spotify functionality within Alexa), unlike in physical space, there are no cues that you are in that "area," nor are there any cues as to what you can do or how you can interact. I can't help but hope that experts in accessibility and sound-based interfaces will save the day and help us to improve today's impressive—but still suboptimal—voice interfaces.

As product and service designers, we're here to solve problems, not come up with new puzzles for our users. We should strive to match our audience's perception of virtual space (whatever that may be) as best we can, and align our offerings to the ways our users already interact with other things or humans. Let's put those parietal lobes to use!

Further Reading

Gallistel, C. R. (1990). *The Organization of Learning*. Cambridge, MA: MIT Press.

Müller, M., & Wehner, R. (1988). "Path Integration in Desert Ants, Cataglyphis Fortis." *Proceedings of the National Academy of Sciences*, 85(14): 5287–5290.

[4]

Memory/Semantics

Abstracting Away the Detail

IT MAY NOT FEEL like it, but as we take in a scene or a conversation, we are continuously dropping a majority of the concrete physical representation of the scene, leaving us with a very abstract and concept-based representation of what we were focusing on. But perhaps you feel like you are much more of a "visual thinker" and really *do* get all the details. Great! Please tell me which of the ones shown in Figure 4-1 is the real US penny.

FIGURE 4-1
Which is the real US penny?

If you are American, you may have seen thousands of these in your lifetime. So surely this isn't hard for a visual thinker! (You can find the answers to this and the following riddle at the end of the chapter.)

OK, that last test might be considered unfair if you aren't American or rarely use physical currency. In that case, let's consider a letter you've seen millions of times: the letter "G". Which of the following is the correct orientation of the letter "G" in lowercase? (See Figure 4-2.)

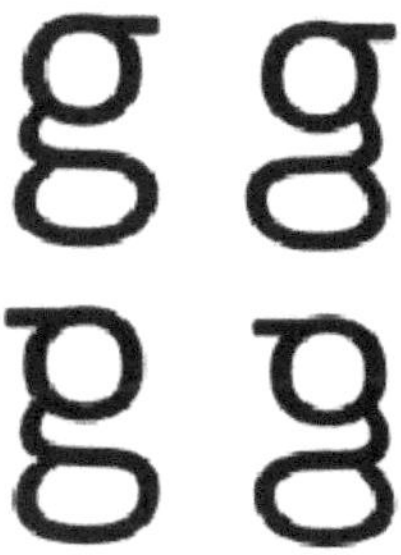

FIGURE 4-2
Which is the real "G"?

Not so easy, right? In most cases, when we look at something, we feel like we have a camera snapshot in our mind. But in less than a second, your mind discards the physical details and reverts to a stereotype or abstract concept of it—and all the assumptions that go along with it.

Remember, not all stereotypes are negative. The actual Merriam-Webster definition is "something conforming to a fixed or general pattern." We have stereotypes for almost anything: a telephone (Figure 4-3), coffee cup, bird, tree, etc.

When we think of these things, our memory summons up certain key characteristics. These concepts are constantly evolving (e.g., from wired telephone to mobile phone). Only the older generations might pick the representation on the right as a "phone."

In terms of cognitive economy, it makes logical sense that we wouldn't store every perspective, color, and light/shadow angle of every phone we have ever seen. Rather, we move quickly from a specific instance of a phone to the concept of a phone. The conceptual representation fills in any gaps in our memory about a specific instance (e.g., the back of the phone that you never actually saw).

FIGURE 4-3
Stereotypes of phone

[SIDE NOTE]

As product designers, we can use this quirk of human cognition to our benefit. By activating an abstract concept that is already in someone's head (e.g., the steps required to buy something online), we can efficiently manage expectations, be consistent with expectations, and make the person more trusting of the experience.

TRASH TALK

Let me provide you with an experiment to show just how abstract our memory can be. Get out a piece of paper and a pencil, and draw an empty square on the piece of paper (or use the space provided in Figure 4-4). Then, after reading this paragraph, go to Figure 4-5 and look at the image for 20 seconds (don't pick up your pencil yet, though). After 20 seconds are up, scroll back or hide the page so you can't look at the image. Then, pick up your pencil and draw everything that you saw.

It doesn't have to be a Rembrandt (or an abstract Picasso), just a quick big-picture depiction of the objects you saw and where they were in the scene. Just a sketch is fine—and you can have 2 minutes for that.

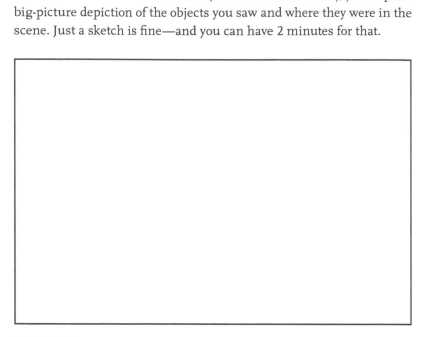

FIGURE 4-4
Draw your image here

OK, go! Remember, 20 seconds to look (no drawing), then 2 minutes to sketch (no peeking).

Since I can't see your drawing (though I'm sure it's quite beautiful), I'll need you to grade it yourself. Look at the original image and compare it to your sketch. Did you capture everything? Two trash cans, one trash can lid, a crumpled-up piece of trash, and the fence?

Good! Now, going one step further, did you capture the fact that one of the trash cans and the fence are both cut off at the top? Or that you can't see the bottom of the trash cans or lid? I didn't think so.

FIGURE 4-5
Picture of an alleyway

When many people see this image, or images like it, they unconsciously "zoom out" and complete the objects according to their stored representations of similar objects. In this example, they tend to extend the fence so its edges go into a point, make the lid into a complete circle, and sketch the unseen edges of the two garbage cans. All of this makes perfect sense if you are using the stereotypes and assumptions you have about trash cans, but it isn't consistent with what you actually saw in this particular image.

Technically, we don't *know* what's actually beyond the rectangular frame of this image. We don't *know for sure* that the trash can lid extends beyond what we can see, or that the fence top ends just beyond what we can see in this image. There could be a whole bunch of statues of David sitting on top of the fence, for all we know (Figure 4-6).

FIGURE 4-6
Did you draw these statues above the tops of the fence posts? (Source: *https://flic.kr/p/4t29M3*)

Our natural tendency to mentally complete the image is called "boundary extension" (Figure 4-7). Our visual system prepares for the rest of the image, as if we were looking through a cardboard tube or a narrow doorway. Boundary extension is just one example of how our minds move quickly from very concrete representations of things to representations that are much more abstract and conceptual.

FIGURE 4-7
Examples of boundary extension

The main implication for product managers and designers is this: a lot of what we do and how we act is based on unseen expectations, stereotypes, and anticipations, rather than what we're actually seeing when light hits the back of our retinas. We as product and service designers need to discover what our audience's hidden anticipations and stereotypes might be (as we'll discuss in Part II of the book).

Stereotypes of Services

Human memory, as we've been discussing, is much more conceptual than people generally think it is. When remembering a scene (e.g., eyewitness testimony), people often forget many perceptual details and rely on what is stored in their semantic memory. The same is true of events. How many times have you heard a parent talk about the time that one of their kids misbehaved many years ago, and incorrectly blame it on "the child that was always getting into trouble" rather than the "good one?" I was fortunate enough to be in the latter camp and got away with all kinds of things according to my mom's memory, thanks to stereotypes.

The trash can drawing shown earlier was a visual example of stereotypes, but they need not be visual. We also have stereotypes about how things might work, and how we might interact in certain situations. Here's an example that has to do more with language, interactions, and events.

Imagine inviting a colleague to a celebratory happy hour. In their mind, "happy hour" may mean swanky decorations, modern bar stools, drinks with fancy ice cube blocks, and sophisticated "mixologists" with impeccable clothing. Happy hour in your mind, on the other hand, might mean sticky floors, $2 beers on tap, and the same grumpy guy named "Buddy" in the same old T-shirt asking "Whatcha want?" See Figure 4-8 for a comparison.

FIGURES 4-8
What is "happy hour" to you?

Both of these are "happy hour," but the underlying expectations of what's going to happen in each of these places might be very different. Just like we did in the sketching exercise, we jump quickly to abstract representations. We anticipate where we might sit, how we might pay, what it might smell like, what we will hear, who we will meet there, how we will order drinks, and so on.

In product and service design, we need to know what concept is associated with a term according to our customers. "Happy hour" is a perfect example. When there is a dramatic difference between a customer's expectation of a product or service and how we designed it, we are suddenly fighting an uphill battle by trying to overcome our audience's well-practiced expectations.

The Value of Understanding Mental Models

Knowing and activating the right mental models—*defined* as "psychological representations of real, hypothetical, or imaginary situations—can save us a huge amount of time as product or service designers. This is something we rarely hear anything about in customer experience—and yet, understanding and activating the right mental models will build trust with our target audience and reduce the need for instructions.

CASE STUDY: THE CONCEPT "WEEKEND"

Challenge: In one project for a financial institution, my team and I interviewed two groups of people regarding how they use, manage, and harness money to accomplish their goals in life. The two groups consisted of: (a) a set of young professionals, most of whom were unmarried, without children; and (b) a group that were a little bit older, most of whom had young children. We asked them what they did on the weekend. You can see their responses in Figure 4-9.

Outcome: Clearly, the two groups had very different semantic associations with the concept of "weekend." Their answers helped us glean: (a) what the word "weekend" means to each of these groups; and (b) how the two groups are categorically different, including in what they value and how they spend their time. Our further research found very large differences in the concept of luxury for each group. In tailoring products/services to each of these groups, we would need to keep in mind their respective mental models of "weekend." This could influence everything from the language and images we use to the emotions we try to evoke.

FIGURE 4-9
Words used to describe a weekend

Acknowledging the Diversity of Types of Mental Models

Thus far, I've discussed how our minds go very quickly from specific visual details or words to abstract concepts, and how the representations that are generated by those visual features or words can be distinct across audiences. But in addition to these perceptual or semantic patterns, there are many other types, such as stereotypical eye patterns and motor movements.

You've probably experienced being handed someone's phone or a remote control you've never used before and saying to yourself something like: "Ugh! Where do I begin? Why is this thing not working? How do I get it to…? I can't find the…" That experience is the collision between your stereotypical eye and/or motor movements, and the need to override them.

The point I'm driving home here is that there are expectations your customers have about interactions with products and services. That we are built this way makes sense because under normal circumstances, stored and automated patterns are very efficient and allow our mental focus to be elsewhere. As product and service managers and designers, we need to:

- Understand our users' many diverse stored concepts and automatic processes

- Anticipate (and counteract) confusion if and when we deviate from those mental assumptions

Riddle Answer Key!

Page 31: Which is the real US penny?

Answer: Row 1, column 4. If you didn't get it right, join the club. Most times, answers are all over the map.

Page 32: Which is the real "G"?

Answer: The one at the top left is the O.G. (Original "G").

Further Reading

Intraub, H., & Richardson, M. (1989). "Wide-Angle Memories of Close-Up Scenes." *Journal of Experimental Psychology: Learning, Memory, and Cognition.* 15(2): 179–187 *http://doi.org/10.1037/0278-7393.15.2.179*

Wong, K., Wadee, F., Ellenblum, G., & McCloskey, M. (2018). "The Devil's in the g-Tails: Deficient Letter-Shape Knowledge and Awareness Despite Massive Visual Experience." *Journal of Experimental Psychology: Human Perception and Performance.* 44(9): 1324–1335 *http://doi.org/10.1037/xhp0000532*

[5]

Language: I Told You So

IN VOLTAIRE'S WORDS, "LANGUAGE is very difficult to put into words." But I'm going to try to anyway.

In this chapter, we're going to discuss what words our audiences are using, and why it's so important for us to understand what those words tell us about how we should design our products and services.

Wait, Didn't We Just Cover This?

In the previous chapter, we discussed our mental representations of meaning. We have linguistic references for these concepts as well. Often, nonlinguists think of a concept and the linguistic references to that concept as one and the same. But they're not. Words are actually strings of morphemes/phonemes/letters that are associated with semantic concepts. Semantics are the abstract concepts that are associated with the words. In English, there is no relationship between the sounds or characters and a concept without the complete set of elements. For example, "rain" and "rail" share three letters, but that doesn't mean their associated meanings are nearly identical. Rather, there are essentially random associations between a group of elements and their underlying meanings (see Figure 5-1).

What's more, these associations can differ from person to person. This chapter focuses on how different subsets of your target audience (e.g., nonexperts and experts) can use very different words, or use the same word but attach different meanings to it. This is why it's so important to carefully study word use to inform product and service design.

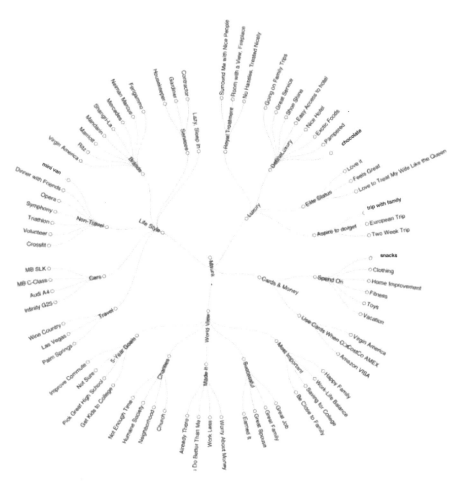

FIGURE 5-1
Semantic map

The Language of the Mind

As humans and product designers, we assume that the words we utter have the same meanings for other people as they do for us. Although that might make our lives, relationships, and designs much easier, it's simply not true. Just like the abstract memories that we looked at in the previous chapter, word–concept associations are more unique across individuals and especially across groups than we might realize. We might all do better in understanding each other by focusing on what makes each of us unique and special.

Because most consumers don't realize this, and have the assumption that "words are words" and mean what they believe them to mean, they are sometimes shocked—and become less trusting—when those products or services use unexpected words or unexpected meanings for words. This can include anything from cultural references ("BAE") to informality in tone ("dude!") to technical jargon ("apraxic dysphasia").

If I told you to "use your noggin," for example, you might accept this advice to try to concentrate harder on something—or you might be offended. If you're a fellow cognitive scientist, you might find the informality of "noggin" insultingly imprecise. If you're not, and I told you instead to "use your dorsolateral prefrontal cortex," you might find my language confusing ("Is that even English?"), meaningless, or scary ("Can I catch that in a public space?"). Either way, I run the risk of losing your trust by deviating from your expected style of prose (see Table 5-1 for more examples).

TABLE 5-1. The terms we use can reveal our depth of understanding

LAYPERSON	COGNITIVE NEUROPSYCHOLOGIST
Stroke	Cerebral vascular accident (CVA)
Brain freeze	Transient ischemic attack (TIA)
Brain area near the middle of your forehead	Anterior cingulate gyrus

The same challenge applies to texting (see Table 5-2). Have you ever received a text reading "SMH" or "ROTFL" and wondered what it meant? Or perhaps you were the one sending it, and received a confused response from an older adult. Differences in culture, age, and geographic location are just a few of the factors that influence the meanings of words in our minds, or even the existence of certain entries in our mental dictionaries—our mental "lexicon."

TABLE 5-2. Young adults are prolific creators of new words

ADULT TEXTER'S TERMS	TEEN TEXTER'S TERMS
I'll be right back	BRB
That's really funny	ROTFL
For what it's worth	FWIW
In my opinion	IMHO

"What We've Got Here Is Failure to Communicate"

B2C communication failures often use over-jargony language that confuses customers, causing them to lose faith in the company and end the relationship. Have you ever seen an incomprehensible error message on your laptop? Or been frustrated with an online registration form that asks you to provide things you've never even heard of (e.g., this actual health care enrollment question: "What is your FBGL, in mg/dl")?

This failure to communicate usually stems from a business-centric perspective, resulting in overly technical language or, sometimes, an overenthusiastic branding strategy that results in the company being too cryptic with its customers (what *is* the difference between a "venti" and a "tall"?). To reach your customers, it's crucial that you understand the customers' level of sophistication in your line of work (as opposed to your intimate in-house knowledge of it), and that you provide products that are meaningful to them at their level.

[SIDE NOTE]

Case in point: Did you catch my *Cool Hand Luke* reference in this section's title? You may or may not have, depending on your level of expertise when it comes to Paul Newman movies from the '60s, or your age, or your upbringing. If I were trying to reach Millennials in a clever marketing campaign, I probably wouldn't quote from that movie; instead, I might choose something from *The Matrix*.

Revealing Words

The words that people use when describing something can reveal their level of expertise. If I'm talking with an insurance agent, for example, they may ask whether I have a PLUP. For that agent, it's a perfectly normal word, even though I may have no idea what a PLUP is (in case you don't either, it's short for a Personal Liability Umbrella Policy, which provides coverage for any liability issue—upon first hearing what the acronym stood for, I thought it might protect you from rain and flooding!).

Over time, people like this insurance agent build up expertise and familiarity with the jargon of their field (see Table 5-3). The language they use suggests their level of expertise. To reach them (or any other potential customer), we need to understand:

- The words people are using

- What meanings are associated with those words

TABLE 5-3. Are laypeople and experts understanding each other?

LAYPERSON'S TERMS	INSURANCE BROKER'S TERMS
Home insurance	Annualization
Car insurance	Ceded Reinsurance Leverage
Liability insurance	Personal Liability Umbrella Policy (PLUP)

As product owners and designers, we want to make sure we're using words that resonate with our audience—words that are neither over nor under their level of expertise. If we were communicating with a group of orthopedic specialists, we would use very different language than if we were trying to communicate to young grade school students. If we tried to use the specialists' complicated language when speaking to children, instead of layman's terms, we'd run the risk of confusing and intimidating our audience, and probably losing their trust as well.

Perhaps this is why *Cancer.gov* provides two definitions of each type of cancer: the health professional version and the patient version. You've heard people say "you're speaking my language." Just like Cancer.gov, you want your customers to experience this same comfort level when they come across your products or services—whether as an expert or a novice. It's a comfort level that comes from a common understanding and leads to a trusting relationship.

When your products and services have a global reach, there is also the question of the accuracy of translation, and the identification and use of localized terms (e.g., in Table 5-4, *Chesterfield* in Canada would be *couch* in the US). We must ensure that the words that are used in each location mean what we want them to mean when they're translated into a different language or dialect.

I remember a Tide detergent ad from several years ago that said things like "Here's how to clean a stain from the garage (e.g., oil), or a workshop stain, or lawn stain." While the translations were reasonably accurate, the intent went awry when it was used in India and Pakistan. Why? Indian and Pakistani populations mostly have "flats" (apartments) without garages, workshops, or lawns. Their conceptual structure was entirely different!

I'm Listening

Remember in the previous chapter, how I used the example of my team using interviews with young professionals and parents of young children to uncover underlying semantic representations among our audience? I can't overstate the importance of interviews, and transcripts of those interviews, in researching your audience. We want to know the exact terms they use (not just our interpretation of what they said) when we ask a question like, "What do you think's going to happen when you buy a car?" If you're a car salesperson, examining transcripts will often reveal that the lexicon your customers use is very different from your own.

Through listening to their exact words, we can learn what *words our customers are commonly using, the level of expertise their words imply*, and ultimately, *what sort of process they're expecting*. This helps experience designers either conform more closely to customers' anticipated experience or warn their customers that the process may differ from what they might expect.

Overall, here's our key takeaway. It's pretty simple, or at least it sounds simple enough: once we have an understanding of our customers' level of understanding, we can create products and services that have the sophistication and terminology that works best for them. This leads to a common understanding and builds trust—ultimately leading to happy, loyal customers.

[6]

Decision Making and Problem Solving: Enter Consciousness, Stage Left

MOST OF THE PROCESSES I've introduced so far, like attentional shifts and associating words with their meanings, occur automatically, even if they're influenced by consciousness. In contrast, this chapter focuses on the very deliberate and conscious process of decision making and problem solving. Relative to other processes, this is the one that you're the most aware of and in control of. Right now as you are reading this, you're aware that you're thinking about thinking and decision making.

In this chapter we will focus on how we, as decision makers, define where we are now and our goal state, and make decisions to get us closer to our desired goal. Designers rarely think in these terms, but I hope to change that.

[SIDE NOTE]

There are several fantastic books and articles about how we as humans deviate from perfect rationality when making decisions—Amos Tversky and Daniel Kahneman won a Nobel Prize demonstrating that. In the next chapter we will examine emotions and some of the ways they factor into decision making. For now, though, let's focus on the decision-making process generally.

What Is My Problem (Definition)?

When you're problem solving and decision making, you have to answer a series of questions. The first one is, "What is my problem?" By this I mean "What is the problem you're trying to solve?" Where are you now (current state), and where do you want to go (goal state)?

For example, consider Escape Room, an adventure game where you have to solve a series of riddles as quickly as possible to get out of a room (Figure 6-1). While unlocking the door may be your ultimate goal, there are subgoals you'll need to accomplish before that (e.g., finding the key) in order to get to your end goal. For example, you may need to open a locked glass cabinet that appears to hold the key, find clues about how to unlock the cabinet, etc. (I'm making these up; no spoiler alerts here!).

FIGURE 6-1
People searching for clues in an escape room

Chess is another example of building subgoals within larger goals. Your ultimate goal is to checkmate the opponent's king. As the game progresses, however, you'll need to create subgoals to help you reach your ultimate goal. Your opponent's king is protected by his queen and a bishop, so a subgoal (to the ultimate goal of putting the opponent's king into checkmate) could be eliminating the bishop. To do this, you may want to use your own bishop, which then necessitates another subgoal of moving a pawn out of the way to free up that piece. Your opponent's moves will also trigger new subgoals for you—such as getting your queen out of a dangerous spot, or transforming a pawn by moving it across the board. In each of these instances, subgoals are necessary to reach the desired end goal.

How Might Problems Be Framed Differently?

Remember when we talked about experts and novices in the last chapter, and the unique words each group uses? When it comes to decision making, experts and novices are often thinking about (or "framing") the problem very differently, too.

Let's consider buying a house, for example. The novice home buyer might be thinking, "What amount of money do we need to offer for the owner to accept our bid?" Experts, however, might be thinking about several more things: Can this buyer qualify for a loan? What is their credit score? Have they had any prior issues with credit? Do the buyers have the cash needed for a down payment? Will this house pass an inspection? What are the likely repairs that will need to be made before the buyers might accept the deal? Is this owner motivated to sell? Is the title of the property free and clear of any liens or other disputes?

So while the novice home buyer might frame the problem as just one challenge (convincing the buyer to sell their house at a specific price), the expert is thinking about many other things as well (e.g., a "clean" title, building inspection, credit scores, seller motivations, etc.). From these different perspectives, the problem definition is very different, and the decisions the two groups make and actions they might take will also likely be very different.

In many cases, novices (e.g., first-time home buyers or car buyers) don't define the problem that they really need to solve because they don't understand all the complexities and decisions they need to make. Their knowledge of the problem might be overly simplistic relative to reality.

This is why the first thing we need to understand as product designers is how our customers define the problem they are solving. We need to meet them there, and, over time, help to redirect them to what their actual (and likely more complex) problem is and help them along the way. This is known as *redefining the problem space*.

◀ 1 of 2 ▶

Breville Fresh & Furious Blender

Sugg. Price $299.95
Our Price $199.95

◀ 1 of 3 ▶

Breville Q Blender

Sugg. Price $549.95
Our Price $349.95

◀ 1 of 5 ▶

Breville Super Q Blender

Sugg. Price $799.95
Our Price $499.95

FIGURE 6-2
Williams Sonoma blenders

MUTILATED CHECKERBOARD PROBLEM

A helpful example of redefining a problem space comes from the so-called "mutilated checkerboard problem," as defined by cognitive psychologists Craig Kaplan and Herbert Simon. The basic premise is this: imagine you have a checkerboard. (If you're in the US, you're probably imagining intermittent red and black squares; if in the UK, you might call this a chessboard, with white and black squares. Either version works in this example.) It's a normal checkerboard, except for the fact that you've removed the two opposite black corner squares from the board, leaving you with 62 squares instead of 64. You also have a bunch of dominoes, which cover two squares each (see Figure 6-3).

FIGURE 6-3
Mutilated checkerboard

Your challenge: Arrange 31 dominoes on the checkerboard such that each domino lies across its own pair of red/black squares (no diagonals).

Moving around the problem space: When you challenge someone with this problem, they inevitably start putting dominoes onto the checkerboard to solve it. Near the end of that process, they inevitably get stuck and try repeating the process. (If you are able to solve the problem without breaking the dominoes, be sure to send me your solution!)

The problem definition challenge: The problem as posed is unsolvable. If every domino has to go on one red and one black square, and there are no longer an equal number of red and black squares on the board (since we removed two black squares and no red squares), the problem can't be solved. To the novice, the definition of the problem, and the way to move around in the problem space, is to lay out all the dominoes and figure it out. Each time they put down a domino they believe they are getting closer to the end goal. They've likely calculated that since there are now 62 squares, and 31 dominoes, and each domino covers 2 spaces, the math works. An expert, however, instantly knows that you need equal numbers of red and black squares to make this work, and won't bother to try to lay dominoes on the checkerboard.

Finding the Yellow Brick Road
to Problem Resolution

I've mentioned moving around in the problem space from your starting state to the goal state. Let's look at that component more closely.

First, it's really important that as product or service designers, we make no assumptions about what the problem space looks like for our customers. As experts in the problem space, we know all the possible moves around that space that can be taken, and it often seems obvious what decisions need to be made and what jobs need to be done. That same problem may look very different to our (less expert) customers.

In games like chess, it's clear to all parties involved what their possible moves are, if not all the consequences of their moves. In other realms, like getting health care, the steps aren't always so clear. As designers of these processes, we need to learn what our audiences see as their "yellow brick road." What do they think the path is that will take them from their beginning state to their goal state? What do they see as the critical decisions to make? The road they're envisioning may be very different from what an expert would envision, or what is even possible. But once we understand their perspective, we can work to create a product or service that serves to gradually morph their novice mental models into a more educated one so that they make better decisions and understand what might be coming next.

When You Get Stuck En Route: Subgoals

We've talked about problem definition for our target audiences, but what about when they get stuck? How do they get around the things that may block them ("blockers")? Many users see their end goal, but what they don't see, and what we product and service designers can help them see, are the subgoals they must have—and the steps, options, and possibilities for solving those subgoals.

One way to get around blockers is through creating subgoals, like those we discussed in the Escape Room example. You realize that you need a certain key to unlock the door. You can see that the key is in a glass box with a padlock on it. Your new subgoal is opening the padlock (to unlock the cabinet to get the key to unlock the door).

We can also think of these subgoals in terms of questions the user needs to answer. To lease a car, the customer will need to answer many subquestions (How old are you?, How is your credit?, Can you afford the monthly payments?, Can you get insurance?) before the ultimate question (Can I lease this car?) can be answered. In service design, we want to address all of these potential subgoals and subquestions for our user so they feel prepared for what they're going to get from us. It's important that we address these microquestions in a logical progression.

Ultimately, you as a product or service designer need to understand:

- The actual steps to solve a problem or make a decision
- What your audience thinks the problem or decision is and how to solve it
- The subgoals your audience might create to get around "blockers"
- How to help the target audience shift their thinking from that of a novice to that of an expert in the field (changing their view of the problem space and subgoals) to be more successful

Now that we have considered decision making and problem solving from a logical, rational, "Does this make sense?" level, in the next chapter we'll consider how emotions and decision making are inherently intertwined.

Further Reading

Ariely, D. (2008). *Predictably Irrational: The Hidden Forces that Shape Our Decisions*. New York: HarperCollins.

Pink, D. H. (2009). *Drive: The Surprising Truth About What Motivates Us*. New York: Riverhead Books.

Thaler, R. & Sunstein, C. (2008). *Nudge: Improving Decisions About Health, Wealth and Happiness*. New York: Penguin Books.

[7]

Emotion: Logical Decision Making Meets Its Match

Up until now, we've treated everyone like they are perfectly rational and make sound decisions every time. While I'm sure that applies in your case (not!), most of us systematically deviate from logic, and often use mental shortcuts. When overwhelmed, we default to heuristics and end up "satisficing"—picking an option that is easy to recall and seems about right, rather than using careful decision making.

As product designers, I want you to think about all the emotions that are critical to product and service design (see Figure 7-1). This does mean considering the emotions and emotional qualities that are evoked as a customer experiences our products and services. But it also means going deeper, to the customer's underlying and deep-seated goals and desires (which I hope you will help them accomplish with your product or service), as well as the customer's biggest fears (which you may need to design around should they play a role in decision making).

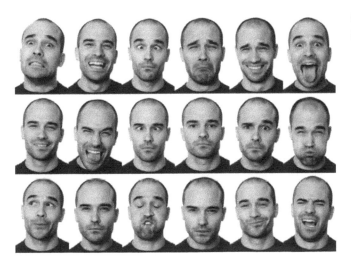

FIGURE 7-1

Portraits of emotion

Too Much Information Jamming Up My Brain!
Too Much Information Driving Me Insane!

I mentioned Daniel Kahneman earlier in reference to his work on attention and mental effort in *Thinking, Fast and Slow*; he shows how, for example, in a quiet room, by yourself, you can usually make quite logical decisions. If, however, you're trying to make that same decision in the middle of a New York City subway platform at rush hour, with someone shouting in the background and your child tugging at your arm, you'll simply be unable to make as good a decision. This is because all of your attention and working memory are being occupied with other things.

Herbert Simon coined the notion of *satisficing*, which means accepting an available (easily recallable) option as not necessarily the ideal decision or choice, but perhaps satisfactory given the limited cognitive resources available for decision making at the time. At times when you are mentally taxed, either due to overstimulation or emotion, you often rely on a gut response—a quick, intuitive association or judgment.

It makes sense, right? Simply having your attention overwhelmed can dramatically affect how you make decisions. If I ask you what 17 minus 9 is, for example, you'll probably get the answer right fairly quickly. If I ask you to remember the letters A-K-G-M-T-L-S-H in that order and be ready to repeat them, and while holding onto those letters ask you to subtract 8 from 17, however, you are likely to make the same arithmetic errors that someone who suffers from math phobia would produce. For those who get extremely distraught and emotional thinking about and dealing with numbers, those worries can fill up their working memory capacity and impair their ability to make rational decisions, forcing them to fall back on strategies like satisficing.

Some businesses have mastered the dark art of getting consumers to make suboptimal decisions. That's why casinos overwhelm you with lighting and music and drinks, and make sure clocks and other time cues are nowhere to be found—to keep you gambling. It's why car dealerships often make you wait around for a while ("Let me check with my manager and see what I can do"), then ask you to make snap decisions through which you either get a car, or nothing. When is the last time a car salesperson asked you to go home and sleep on a deal? I encourage you to do exactly that, to ensure the emotional content is not affecting your decision making.

Spock, I Am Not

With a better understanding of decision making, you might assume that those who study decision making for a living (e.g., psychologists and behavioral scientists) might make more logical, rational decisions, like Captain Kirk's stoic counterpart Spock. Like other humans, we have our rational systems competing with our feelings and emotions as we make decisions. Beyond the cerebral cortex lie more primitive centers that generate competing urges to follow our emotional response and ignore the logical.

Early cognitive psychologists thought about decision making in simple terms, focusing on all of the "minds" you've seen up until now, like perception, semantics, and problem solving. But they left out one crucial piece: emotion. In his 1996 book, "The Emotional Brain," Joseph LeDoux argued that traditional cognitive psychology was making things unrealistically simple. There are so many ways that we deviate from logic, and so many ways that our lower reptilian brains affect our decision making. Dan Ariely demonstrates several of these in his book *Predictably Irrational: The Hidden Forces That Shape Our Decisions*.

This affects us in a myriad of ways. For example, it has been well demonstrated that humans hate losses more than we love gains. "People tend to be risk averse in the domain of gains and risk seeking in the domain of losses," Ariely writes. Because we find more pain in losing than we find pleasure in winning, we don't work rationally in economic and other decisions. To intuitively understand this, consider a lottery. You are unlikely to buy a $1 ticket with a possible payoff of $2. You would want the chance to win $10,000, or $100,000, just from that one ticket. You are imagining what it would be like to have all that money (a very emotional response), just as picturing losing that $1 and not winning can elicit the feeling of loss.

Our irrationality, however, is predictable, as Ariely demonstrates. He argues that we are systematic in the ways we deviate from what would be logically right. According to Ariely, "we consistently overpay, underestimate, and procrastinate. Yet these misguided behaviors are neither random nor senseless. They're systematic and predictable—making us predictably irrational."

Competing for Conscious Attention

Sometimes, your brain is overwhelmed by your setting, like the subway platform example. Other times, it's overwhelmed by emotions.

A good deal of research has gone into all of these systematic deviations from logic, which I simply don't have time to present in this book. But the key point is that whereas in optimal conditions (no time pressure, quiet room, time to focus, no additional stress), we can make great, logical decisions, in the real world, we often lack the ability to concentrate sufficiently to make logical decisions. What we do instead is *satisfice*—we make decisions using shortcuts. An example of a tool we use in lieu of careful thought is: "If I think of a prototypical example of this, does the ideal in my mind's eye match a choice I've been given?"

Imagine yourself in that car dealership negotiating a price. Your two children were as good as gold during the test drive, but they're getting restless and you're growing worried that they are going to fall off a chair or knock something over. You are hungry and tired. The salesperson leaves for what seems like an eternity and finally returns with an offer, which has many lines and includes decisions about percentages down, loan rates, options, damage protection, services, insurance, and much more. During the explanation it happens—one of your children falls and is now crying and talking to you. As you hold their fidgeting body and attempt to listen to the salesperson, you simply don't have the attentional resources to give to the problem at hand (determining if this is a fair deal and which options you want to choose). Instead, you imagine yourself driving on the open road with the sunroof open (far from the car dealership and family)—and that emotional side takes over.

As product designers, we need to understand both what the rational, conscious part of the customer's mind is seeking (data to make good, logical decisions) and what the underlying emotional drivers are for making the decision. It is my hope that you will provide your target audience with the information they need and support them in making the best decision for them, rather than seeking to overwhelm and obfuscate in order to drive emotional decision making. Both rational and emotional responses are crucial in every decision we make. This is why people who are not salespeople often encourage you to "sleep on it" when you have to make a choice, giving you the time you need to make a more informed, less emotional decision.

All of these feelings flooding in are subconscious emotional qualities. Just like when you have your attention overwhelmed on the subway, you have less of your memory available to make good decisions when you're overwhelmed with emotion. (That's why, as a psychologist, I never let a salesperson sit me in a car that I'm not planning to buy. Seriously, don't try me.) We all have emotions competing for our conscious resources. When the competition ramps up, that's how we start to make decisions we regret later.

Getting to Deep Desires, Goals, and Fears

In a market research study for a client in the financial industry, I started out by asking consumers innocuous questions about their favorite credit cards, then asked questions that got progressively deeper as I probed: questions like "What are your goals for the next three years?" and "What worries or excites you most about the future?" The sessions frequently ended in tears and hugs, with respondents saying this was the best therapy session they'd had in a long time. In a series of eight questions, I went from people saying what cards were in their wallets to sharing their deepest hopes and fears. By listening to them, I was able to draw out:

- What would appeal to them immediately

- What would enhance their lives and provide more lasting and meaningful value

- What their deepest goals and wishes were for life

The first two points are essential to getting to the third— but once you get there, you've got your selling point: the deep, underlying meaning of what it is your product is trying to address for your target audience. This is why many commercials don't actually feature the products themselves until the very end, if at all. Instead, they focus on the feeling or image that their customer is trying to mirror: successful businesswoman, family man, thrill-seeking "I've still got it" retiree, etc.

By uncovering (and leveraging) what appeals to your audience immediately, what will help them in the long term, and what will ultimately awaken some of their deepest goals in life, you've gone from the surface level to their gut reaction level—which can't be overestimated in the decision-making process. In the next part of the book we'll see how to get there.

Further Reading

Kahneman, D. (2011). *Thinking, Fast and Slow.* New York: Macmillan.

LeDoux, J. E. (1996). *The Emotional Brain: The Mysterious Underpinnings of Emotional Life.* New York: Simon & Schuster.

Simon, H. A. (1956). "Rational Choice and the Structure of the Environment." *Psychological Review* 63(2): 129–138.

Part II

Exposing Secrets

IF YOU'VE MADE IT this far, chances are you've learned a lot about the six types of mental processes I'd like you to focus on. As a reminder, here are the things you should think about with regard to your target customers:

Vision/Attention

> What is attracting your customers' attention? What are the words, images, and objects they are looking for?

Wayfinding

> How are customers representing where they are (whether in the physical world, an app, or virtual space)? How do they believe they can interact with and navigate in this space?

Memory

> What are the past experiences customers are using to frame and understand what they are experiencing? What are mental models/ stereotypes that are forming their expectations about how things should work and what happens next?

Language

> What are the words your customers are using? What do those words, and their associated meanings according to the customers, say about their level of expertise (thereby suggesting how they might want to be engaged by you)?

Decision Making

> What is the problem your customers think they need to solve? How does that differ from the actual problem? How do they think they can get to the solution? What subproblems do they need solve and what decisions do they need to make along the way?

Emotion

> What are your customers' deep-seated goals, desires, and fears? How are those affecting their decisions, and what they are looking to achieve? How might that affect what will appeal to them?

You may be asking: as a product manager or product or service designer without formal training in the psychological sciences, how can I learn about all of these cognitive processes? Will we have the time or budget to understand all these things? Do I really need to know all this to design my product/service?

The answers, I believe, are all positive. Nonpsychologists can learn about these individual cognitive processes through a combination of watching your customers in action (I call it *contextual inquiry*) and interviewing them.

Further, I would argue that you can get all the information you need through qualitative research and watching people work without fancy equipment, huge budgets, or lengthy studies. I am talking about weeks, and not months (or, for our large enterprise and government colleagues, not years!)

The number one reason projects go over budget and take longer than planned is due to changes made in the late stages of production or right after launch because the features needed are different than those that were built. An intimate understanding of your customers will dramatically reduce the chance that your product misses the mark and requires timely and costly refactoring.

This part of the book lays out how you and your team, whether trained in research or not, can extract the information you need for product and service design by watching your audience and interviewing them. I've met great designers that do much of this instinctively, but it took them years to hone their skills. Why not go from good to great faster and without so much trial and error? Let's get started!

[8]

User Research: Contextual Interviews

MARKET RESEARCH HAS TAKEN many forms through the years. Some may immediately think of the kind of focus group shown in the TV show *Mad Men*. Others may think of large surveys, and still others may have conducted empathy research when using a *design thinking* approach to product and service design, which I'll define later in this chapter.

While focus groups, surveys, and empathy interviews can be great tools to get to what people are *saying*, and maybe some of what they are *doing*, they don't get to the *why* behind these behaviors. Nor do they get us the level of detail in analysis we would like to have to meaningfully influence product and service design decisions.

In this chapter, I'll recommend a different take on market research that combines watching and interviewing people in their typical work or play, and interviewing them. If you've done some qualitative studies before, you might have a fair bit of interesting data to work with already. And if you don't have that data, collecting it is within your grasp. What I'm proposing is designed so that anyone can conduct the research—no psychology PhDs or white lab coats required. It may be very familiar to my UX, psychology, and anthropology readers: it's called a *contextual interview*.

Why Choose a Contextual Interview?

If I had to get to the essence of what a contextual interview is, I'd say "looking over someone's shoulder and asking questions," with a focus on observing customers where they do their work (e.g., at their desks at the office, or at the checkout counter) or where they live and play.

The number one reason digital products take longer getting to market and cost more than planned is a mismatch between user needs and functionality. We need to know what our customers' needs are. Unfortunately, we can't learn what we need simply asking by them. There are several reasons why this is the case.

First, customers often just want to keep doing what they're doing, but better. As product and service designers who are outside that day-to-day grind, we can sometimes envision possibilities beyond today's status quo and leapfrog to a completely different, more efficient, or more pleasurable paradigm. It is not a customer's job to envision what's possible in the future; it's ours!

Second, there are a lot of nuanced behaviors people do unconsciously. When we watch people work or play in the moment, we can see some of the problems with an experience or identify things that don't make sense—that customers compensate for without realizing it. How likely is it that customers are going to be able to report behaviors they themselves aren't even aware of?

For example, I've observed Millennials flipping wildly between apps to connect with their peers socially. They never reported flipping back and forth between apps, and I don't think they were always conscious of what they were doing. Without actively observing them in the moment, I might never have known about this behavior, which turned out to be critical to the products my team was designing.

We also want to see the amazing lengths to which "superusers"—users of your products and services who really need them—go in order to make existing (flawed) systems work. We'll talk about this later, but this notion of watching people "in the moment" is similar to what those in the Lean Startup movement call GOOB, or Getting Out Of (the) Building, to truly see the context in which your users are living.

Third, if your customers are not "in the moment," they often forget all the important details that are critical to creating successful product and service experiences. Memory is highly contextual. For example, I am confident that when you visit somewhere you haven't been in years (e.g., your old elementary school), you will remember things about your childhood that you wouldn't have otherwise because the context triggers those memories. The same is true of customers and their recollections.

In psychology or anthropology circles, watching people to learn how they work is not a new idea at all. Corporations are starting to catch on; it's becoming more common for companies to have a "research anthropologist" on their staff who studies how people are living, communicating, and working. (Fun fact: There is even one researcher that calls herself a cyborg anthropologist! Given how much we rely on our mobile devices, perhaps we all are cyborgs, and we all practice cyborg anthropology!)

Jan Chipchase, founder and director of human behavioral research group Studio D, brought prominence to the anthropological side of research through his research for Nokia. Through in-person investigation, which he calls "stalking with permission" (see, it's not just me!), he discovered an ingenious and off-the-grid banking system that Ugandans had created for sharing mobile phones.

> I never could have designed something as elegant and as totally in tune with the local conditions as this. ... If we're smart, we'll look at [these innovations] that are going on, and we'll figure out a way to enable them to inform and infuse both what we design and how we design.
>
> **JAN CHIPCHASE, "THE ANTHROPOLOGY OF MOBILE PHONES," TED TALK, MARCH 2007**

Chipchase's approach uses classic anthropology as a tool for building products and thinking from a business perspective. I'll now explain how you can do this, too.

Empathy Research: Understanding What the User Really Needs

Chipchase's work is just one example of how we can only understand what the user really needs through stepping into their shoes—or ideally, their minds—for a little while.

LEAVE ASSUMPTIONS AT THE DOOR AND EMBRACE ANOTHER'S REALITY

You need to start by dispelling your (and your company's) assumptions about what your customers need and begin thinking like them instead. In its Human-Centered Design Toolkit, IDEO writes that the first step to design thinking is empathy research, or a "deep understanding of the problems and realities of the people you are designing for."

In my own work, I've been immersed in the worlds of people who create new drugs, traders managing billion-dollar funds, organic goat farmers, YouTube video stars, and people who need to buy many millions of dollars' worth of "shotcrete" (like concrete, but it can be pumped) to build skyscrapers. Over and over, I've found that the more I'm able to *think* like these people, the better I'm able to identify opportunities and optimize product and service designs.

Suppose, however, that you *were* the customer in the past (or worse yet, your boss was the customer decades ago). You and/or your boss might think you know exactly what customers want and need, making research unnecessary. Wrong! You are not the customer, and when you do research in this context it can make it even harder because you must fight against preconceived notions to be able to listen to customers about their needs today.

I remember one client who in the past had been the target customer for his products—before the advent of smartphones. Imagine being at a construction site and purchasing concrete 10 years ago, around the time of flip phones (if you were lucky!). The world has changed so much since then, and surely the way we purchase concrete has, too. This is why it is important to park your expectations at the door, embrace your customer's reality, and live today's challenges.

Here's just one example I noticed while securing a moving truck permit. To give me the permit, the government employee I worked with had to walk to one end of this huge office to get a form, walk all the way over to the opposite corner to stamp it with an official seal, walk nearly as far over to the place where they could photocopy it, and then bring it to me. Meanwhile, the line behind me got longer and longer. Seeing this inefficient process left me wondering why these three items weren't grouped together. It's an example of the unexpected improvements you can discover just through watching people at work. I'm not sure the government worker even noticed the inefficiencies!

Moments like these abound in our everyday lives. Stop and think for a second about a clunky system you've witnessed. Perhaps it was the payment system on the subway? Your health care portal? An app? What could have made the process smoother for you? Once you start observing, you'll find it hard to stop. Trust me. Your kids, friends, and relatives will need to be patient with you and your "helpful hints" from here on out!

ANY INTERVIEW CAN BE CONTEXTUAL

Because so much of memory is contextual and there are so many things our customers will do that are unconscious, we can learn a great deal we wouldn't otherwise when we are immersed in their worlds. That means meeting with farmers in the middle of Pennsylvania, sitting with traders in front of their big banks of screens on Wall Street, having happy hour with high-net-worth jet-setters near the ocean (darn!), observing folks who do tax research in their windowless offices, or even chatting with Millennials at their organic avocado toast joints. The key point is, they're all doing what they normally do.

Contextual interviews allow the researcher to see workers' Post-its on their desks, what piles of paper are active and what piles are gathering dust, how many times they're interrupted, and what kind of processes they actually follow (which are often different than the ones that they might describe during a traditional interview). Your product or service has to be useful and delightful for your customer, which means you need to observe them and how they work. The more immersive and closer to their actual day this experience is, the better (Figure 8-1).

FIGURE 8-1
Observing how a small business owner is organizing their business

In contextual interviews, while I want to be quiet sometimes and just observe, I also ask my research subjects questions like:

- What would I need to know to be successful at your job (e.g., if I took over when you were sick)?

- Where would I get started?

- What would I have to keep in mind?

- What could go wrong?

- What drives you crazy sometimes?

WHAT RESEARCHERS NOTICE

Researchers who do contextual interviews typically consider the following:

Artifacts

What's on the desk (Figure 8-2)? What papers, files, spreadsheets, etc. does this person use to keep track of everything? What else is nearby in the environment?

Communication

How is work communicated or reviewed (email, software, discussion, etc.)? How many other people are working with the customer?

Interruptions

What are the interruptions in the customer's job? How often are these? Do they frequently need to move around? What's the noise level like? How many times is their mobile phone interrupting them? Are they constantly hearing loud overhead announcements about the Dow Jones index? (This last example was actually the case for some stock traders I observed. Their workplace was already so loud and stressful that what they needed from my client was essentially an "easy button" that was simple and made their jobs easier.)

Related factors

What other jobs does this person do in addition to the one you are officially observing? How many programs do they have to use on their computer? Are they always on the computer? Are they using their mobile phone?

FIGURE 8-2

Example of the desk of a research participant (why do they have the psychology book *Influence* right in the middle, I wonder?)

WHY NOT SURVEYS OR USABILITY TEST FINDINGS? DISCOVERING THE WHAT VERSUS THE WHY

Clients sometimes assure me that their user research is solid and they need no other data because they've received thousands of survey responses. It is true that this data will give the client an accurate reading of the *what* of the immediate problem (e.g., the customer wants a faster system, step three of a process is problematic, or the mobile app is cumbersome). But as product and service designers, we need to get to the *why* of the problem: the underlying reasons and rationale behind the *what*.

It could be that the customer is overwhelmed by the appearance of an interface, or was expecting something different, or is confused by the language you're using. It could be a hundred different things. It is extremely hard to infer the underlying root cause of an issue from a survey or by talking to your colleagues who built the product or service. We can only determine *why* customers are thinking the way they are by meeting them and observing them in context.

As an example, take a look at the usability test findings in Figure 8-3. Can you tell why these participants are having trouble "Navigating from Code" in the fourth set of bars? (Me neither!) Classic usability test findings often provide the same *what* information. They will tell you that your users were good at some tasks and bad at others, but often won't provide the clues you need to get to the *why*. This is where conducting research using the Six Minds comes to the rescue.

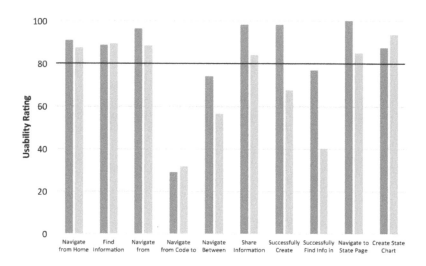

FIGURE 8-3
Usability test findings by task

Recommended Approach for Contextual Interviews and Their Analysis

As I've implied throughout this chapter, shadowing people in the context of their actual work allows you to observe explicit behaviors as well as implicit nuances that your interviewees may not even be aware of. The more that users show you their processes step by step, the more accurate they will be when it comes to reacalling those processes.

With our Six Minds of Experience, I want you to not just experience the situation in context, but also be actively thinking about the many different types of mental representations within your customer's mind:

Vision/Attention

> What are they paying attention to? What are they searching for? Why?

Wayfinding

How are they navigating their existing products and services? How do they believe they should interact with them?

Language

What words do they use? What does that suggest about their level of expertise?

Memory

What assumptions are they making about how things should work? When are they surprised and confused?

Decision Making

What do they say they are trying to accomplish? What does that say about how they are framing the problem? What decisions are they making? What "blockers" are getting in the way?

Emotion

What are their goals? What are they worried about? How might future products or services be better suited to their needs, expectations, desires, and goals?

The sort of observation I've described so far has mostly focused on how people work, but it can work equally well in the consumer space. Depending on what your end product or service is, your research might include observing a family watching TV at home (with their permission, of course), going shopping at the mall with them, or enjoying happy hour or a coffee with their friends. Trust me: you'll have many stories to tell about all the things your customers do that you didn't expect when you return to the office!

This is probably the most fun part, and shouldn't be creepy if you do it correctly. I give you permission (and the participants should provide their or their parents' written permission) to be nosy and curious, and suggest you really question all your existing assumptions. When I'm hiring someone, I often ask them if they like to go to an outdoor cafe and just people-watch. Because that's my kind of researcher: we're totally fascinated by what people are thinking, what they're doing, and why. Why is that person here? Why are they dressed the way they are? Where are they going next? What are they thinking about? What makes them tick? What would make them laugh?

There are some terrific whole books dedicated to contextual interviews, which I'll mention at the end of this chapter. I'll leave it to them to provide all the nuances about these interviews, but I definitely want you to go into your meetings with the following mindset:

You are there to learn and observe

That means you need to blend in, not be the star attraction (your customer is), ask open-ended questions, and leave your assumptions, perspectives, and opinions at the door. Imagine that you are an actor learning to play their role, or that you will be taking over for your customer during maternity/paternity leave—you would never tell them they're wrong, or show them how to do something. You want to learn how to do things their way and think like them.

Follow their customs

Try to dress in a way that is commensurate with your audience, so you are neither intimidating nor stand out. Your goal is to blend in and not influence the situation. If they take off their shoes at the door, so should you. Be ready to sit on the floor, or eat pizza straight from the box.

Try to adopt their language

In other words, try not to be more technical, even if you know more about a subject. Try not to use your company's in-house jargon. In fact, do the opposite—ask the customers what they would call a concept or action and use those terms before you use your words.

Ask why

While sometimes people rationalize and have implicit reasons for an activity, it is always interesting to see what they come up with. Often it helps to understand *how* they are framing a problem or decision, and get clues about their underlying assumptions.

Try to minimize your influence on their actions

It is very hard for someone who is building a product or service not to demonstrate, teach, or promote when they know their product has the perfect feature that could help. That is not your role. (At least, not yet!) You need to observe the customer's perspective, no matter how tough it may be for you to watch. You need to know the reality on the ground.

Make sure you observe them in action

Many times, if you're meeting someone at an office, they'll want to meet in a conference room. While perhaps that would be more comfortable, it's preferable to huddle around their desk and observe them in action in their native environment. Crucially, you want to see them doing the things you seek to improve through your products or services.

Only bring a few people to the contextual interview

One to three is an ideal number. Try to have a small footprint. You do not want your customers to feel like they have an audience, and have to perform for the group. And you don't want the size of the group following them around to constrian their normal activities.

Record information in subtle ways

Do I love to get video and audio recordings? Absolutely! Do I bring along lights and boom poles and fancy mics? No! Bring a wireless mic that the person will forget they are wearing, a compact prosumer video camera, and your cell phone (for pictures). It is a good sign when someone answers the telephone or walks out of the office to ask a colleague a question, because they are following their normal routine and not being polite to you.

Bring a notebook, not a computer

You want to write notes immediately and not worry about time booting up or connecting to WiFi. And I say this from experience: bring an extra pen! One of your colleagues from the office who is following along will forget.

Ask them about themselves, and their perspective

How long have you been doing this? How did you get started? What do you like about the job? What do you do when you are not here? What do you hope to achieve? What makes you happiest? As the observer, you want to be able to see the world through your participants' eyes and understand what makes them tick. Start with normal, socially acceptable questions (e.g., How did you get started in this job?) and gradually get to deeper questions (e.g., What is most important to you in life? What would make you feel accomplished/happy/satisfied?).

Common Questions

Here are some questions that often come up when people are getting started with contextual interviews:

How many people do I need to interview?

Generally I try to estimate the appropriate number of user groups according to their lifestyle or role (e.g., I might interview middle schoolers, high schoolers, and college undergraduates in an academic setting, or in a medical setting I might want to interview general practitioners, specialists, nurses, and administrators). You want about 8 to 12 people per group to identify trends. However, if reality and the ideal clash, remember that any amount is always better than zero!

How long should the contextual interviews be?

I encourage 90-minute interviews. Young kids might not last that long, and busy doctors may refuse that duration. In other cases, you can "ride along" for longer periods (e.g., a morning or afternoon). You want a long enough duration to observe participants' typical pattern of activities *and* talk to them about themselves and their perspective.

How do I recruit participants?

I encourage you to use a professional recruiting service. The time it takes to schedule, remind, reschedule, discuss, prescreen, etc., is far more than you'd imagine if you have never done it before. Recruiters can be well worth the money (and freedom from aggravation). If you choose to recruit on your own and are looking for professionals, start with associations. For the general public, you'd be surprised how far networking and Craigslist can get you. However, when you are flying around the country to conduct interviews, the cost of professional recruitment can be well worth avoiding the situation of getting there and having no one to interview.

Should I have a set of questions prepared?

Yes, but... I encourage you to "go with the flow" of the interviewee. There is a balance to be struck between having participants stay on task and not taking them too far from their standard way of working or living. Don't think of a contextual interview as something for which you need to fill out a form and have every blank filled. Rather, think of it as getting the information you need to step into their world. Often you'll find the conversation heads naturally to what they know, how they frame a problem, and what they value in life. There will be different types of people, and thus it is important to have multiple interviews to define customer segments.

From Data to Insights

Many people get stuck at this step. They have interviewed a set of customers and feel overwhelmed by all of their findings, quotes, images, and videos. Is it really possible to learn what we need to know just through these observations? All these nuanced observations that you've gathered can be overwhelming if not organized correctly. Where should you begin?

To distill hundreds of data points into valuable insights on how you should shape your product or service, you need to identify patterns and trends. To do so, you need the right organizational pattern. Here's what my process looks like.

STEP 1: REVIEW AND WRITE DOWN OBSERVATIONS

In reviewing my own notes and video recordings, I'll pull out bite-sized quotes and insights on users' actions (aka, my "findings"). I write these onto Post-it notes (or on virtual stickies in a tool like Mural or RealTimeBoard). What counts as an observation? Anything that might be relevant to our Six Minds:

Vision/Attention
What are they paying attention to?

Wayfinding
What is their perception of space, and how to move around in that space?

Memory
What are their perspectives on the world?

Language
What are the words they use?

Decision Making
How are they framing the problem? What are they really trying to solve (deeper need)? What "blockers" are getting in the way?

Emotion

What are they worried about? What are their biggest goals?

In addition, if there are social interactions that are important (e.g., how the boss works with employees), I'll write those down as well (Figure 8-4).

FIGURE 8-4

Example of findings from contextual interviews

STEP 2: ORGANIZE EACH PARTICIPANT'S FINDINGS INTO THE SIX MINDS

After doing this for each of my participants, I place all the sticky notes up on a wall, organized by participant. I then align them into six columns, one for each of the Six Minds (Figure 8-5). A comment like "Can't find the 'save for later' feature" might be placed in the Vision/Attention column, whereas "Wants to know right away if this site accepts PayPal" might be filed as Decision Making. Much more detail on this will follow in the next few chapters.

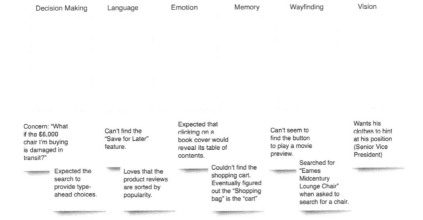

Decision Making	Language	Emotion	Memory	Wayfinding	Vision
Concern: "What if the £5,000 chair I'm buying is damaged in transit?"	Can't find the "Save for Later" feature.	Expected that clicking on a book cover would reveal its table of contents.	Can't seem to find the button to play a movie preview.		Wants his clothes to hint at his position (Senior Vice President)
				Searched for "Eames Midcentury Lounge Chair" when asked to search for a chair.	
Expected the search to provide type-ahead choices.	Loves that the product reviews are sorted by popularity.	Couldn't find the shopping cart. Eventually figured out the "Shopping bag" is the "cart"			

FIGURE 8-5

Getting ready to organize findings into the Six Minds

If you try out this method, you'll inevitably find some overlap; this is completely to be expected. To make this exercise useful for you, however, I'd like you to determine what the *most important* component of an insight is for you as the designer, and categorize it as such. Is the biggest problem visual design? Interaction design? Is the language sophisticated enough? Is it the right frame of reference? Are you providing people with the right tools to solve the problems that they encounter along the way to a bigger decision? Are you upsetting them in some way?

STEP 3: LOOK FOR TRENDS ACROSS PARTICIPANTS AND CREATE AN AUDIENCE SEGMENTATION

In Part III of this book we will talk about audience segmentation. If you look for trends across groups of participants, you'll observe trends and commonalities across findings, which can provide important insights about the future direction for your products and services. Separating your findings into the Six Minds can also help you manage product improvements. You can give the decision-making feedback to your UI expert who worked on the flow, the vision/attention feedback to your graphic designer, and so on. The end result will be a better experience for your users.

In the next few chapters, I'll give some concrete examples from real participants I observed in an ecommerce study. I want you to be able to identify what might count as an interesting data point and to think about some of the nuances that you can get from the insights you collect.

Exercise

In my online classes on the Six Minds, I provide participants with a small set of data I've appropriated from actual research participants (though somewhat fictionalized so I don't give away trade secrets).

In Figures 8-8 through 8-13 are the notes from six participants in an ecommerce research study. They were asked to make purchasing decisions and were either seeking a favorite item, or selecting an online movie for purchase and viewing. The focus of the study was on searching for and selecting the item (the checkout was not a focus of the study). The following notes reflect the findings collected during contextual interviews.

Challenge: Please put each of the notes about the study in the most appropriate category in Figure 8-7 (Vision, Wayfinding, etc.).

Feeling stuck? Perhaps Figure 8-6 can help.

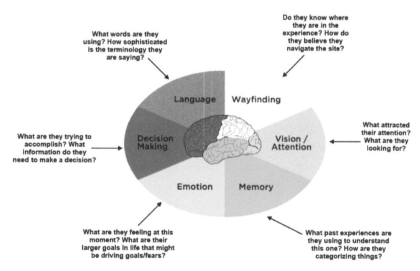

FIGURE 8-6

The Six Minds of Experience

If you feel a note should be put in more than one category, you may do so, but try to limit yourself to the most important category. What did you learn about how each individual was thinking? Were there any trends between participants?

Participant:_____

Decision Making	Language	Emotion	Memory	Wayfinding	Vision

FIGURE 8-7

In which of the Six Minds categories would you place each finding?

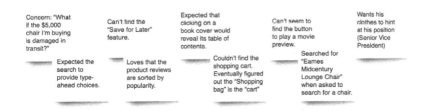

FIGURE 8-8

Findings from participant 1

FIGURE 8-9

Findings from participant 2

FIGURE 8-10

Findings from participant 3

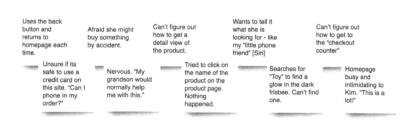

FIGURE 8-11

Findings from participant 4

FIGURE 8-12

Findings from participant 5

I'll return to these five participants and provide snippets of that dat set as needed in the following chapters to concretely illustrate some nuances and help you sharpen your analytic swords so that you know how to handle data in different situations.

Can't wait to complete the exercise and share it with your friends? Great! Please download the Apple Keynote or Microsoft PowerPoint versions to make it easy to complete and share (*http://bit.ly/six-minds-exercise*).

Concrete Recommendations

- Watch people work in their place of work, rather than just interviewing them (many more contextual memories surface when they're "in the moment")

- Watch people complete tasks, not just discuss needs (again, more contextual memories and unconscious behaviors surface this way)

- Try to deduce assumptions about the world that underlie their behavior (allowing yourself to think like the consumer helps you to discover their pain points and issues that you can help with)

- Measure their observable behavior, not just what they say about a topic (How many times are they checking their phone? How many times do they use paper versus. computer?).

Further Reading

Chipchase, J. (2007). "The Anthropology of Mobile Phones" TED Talk. Retrieved January 15, 2019, from *http://bit.ly/2Uy9J1A*.

Chipchase, J., Lee, P., & Maurer, B. (2011). Mobile Money: Afghanistan. *Innovations: Technology, Governance, Globalization*. 6(2): 13–33.

IDEO.org. (2015). "The Field Guide to Human-Centered Design." Retrieved January 15, 2019, from *http://www.designkit.org//resources/1*.

[9]

Vision: Are You Looking at Me?

NOW THAT WE'VE DISCUSSED how to conduct contextual interviews and observe people as they're interacting with a product or service, I want you to think about how those interviews can provide important clues for each of the Six Minds.

I'd like to start by looking at this from a vision/attention perspective (Figure 9-1). In considering vision, we're seeking to answer these questions about their customers:

- Where are their eyes looking? (Where did they focus? What drew their attention? What does that tell us about what they were seeking, and why?)

- Did they find what they were looking for? If not, why? What were the challenges in them finding it?

- What are the ways that new designs might draw their attention to what they're seeking?

We'll discuss not only where customers look and what they expect to see when they look there, but also what this data suggests about what is visually salient to them. We'll consider whether users are finding what they are hoping to find, what their frame of reference is, and what their goals might be.

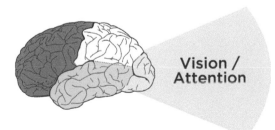

FIGURE 9-1

Vision and visual attention in the Occipital cortex

Vision / Attention

Where Are Their Eyes? Eye-Tracking Can Tell You Some Things, But Not Everything

When it comes to improving interfaces or services, we start with where participants are actually looking. If we're talking about an interface, where are users looking on the screen? Or where are they looking within an app?

Eye tracking devices and digital heat maps come in handy for this type of analysis, helping us see where our users are looking. This sort of analysis can help us adjust the placement of our content on a page.

But you don't always need eye-tracking if you use good old-fashioned observation methods like those we discussed in the previous chapter. When I'm conducting a contextual interview, I try to set myself up at 90 degrees to the participant (so that I'm a little bit behind them without creeping them out, as seen in Figure 9-2). There are a few reasons for this:

- It's a little awkward for them to look over and talk to me. This means that they are primarily looking at the screen or whatever they're doing, and not me (allowing me to better see what it is they're working on, clicking on, etc.).

- I can see what they're looking at. Not one hundred percent, of course, but generally, I'm able to see if they're looking at the top or bottom of the screen, or down at a piece of paper, flipping through a binder to a particular page, etc.

FIGURE 9-2
Moderating a contextual interview

Speaking of where people's eyes are, take a look at Figure 9-3. This figure shows two screens side by side from an electronics company—blurred out a bit, with the color toned down. This is the type of representation your visual system uses to determine where to look next.

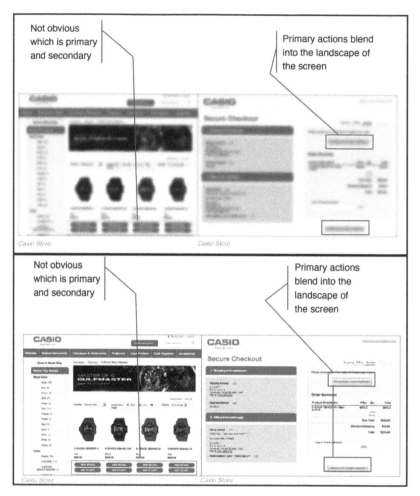

FIGURE 9-3

What your visual attention system sees when you look at an image

In the image on the left there are four watches, with two buttons below each watch. Though you can tell at a glance that these are buttons, it's not clear from the visual features and this level of representation which is the "buy" button and which is the "save for later" button. The latter should appear as a secondary button, yet it currently draws an equal amount of attention as the "buy" button. That's something we would work with a graphic designer to adjust.

In the image on the right, the buttons don't have sufficient visual contrast to draw attention to them, and simply blend into the background, making them easily overlooked.

CASE STUDY: SECURITY DEPARTMENT

Challenge: Even though many of my examples are of digital interfaces, we as designers also need to be thinking about attention more broadly. In this case, I worked with a group of people with an enormous responsibility: monitoring security for a football stadium–sized organization (and/or an actual stadium).

Their attention was divided in so many ways. Here are all the systems and tools (along with their respective alerts, bells, and beeping sounds) they monitored at any given time:

- Literally hundreds of cameras, with the number increasing all the time
- Special cameras focused on problem areas (e.g., the door people frequently used to sneak in the back)
- Walkie-talkies with updates from foot patrols
- Emails
- Texts
- The local police radio (with constant communication)
- Card-swipe systems for badge-controlled doors (which beeped many times a minute)
- Broadcast PA systems
- CNN
- Fire alarms
- Elevator alarms
- Electrical alarms
- Phone banks

If you're impressed that anyone could get work done in such a busy environment, you're not alone; I was shocked (and a bit skeptical about whether all these noisy systems were helping rather than hurting their productivity). Here was an amazing challenge of divided attention, far more distracting than an open office layout (which many people find distracting).

Recommendation: With huge visual and auditory distractions in play, we had to distill the most important thing that staff should be attending to at each moment. My team developed a system very similar to a scroll-based Facebook news feed, except with extreme filtering to ensure the relevancy of the feed (no cat memes here!). Each potential concern (terror, fire, door jams, etc.) had its own chain of action items associated with it, and staff could filter each issue by location. The system also included a prominent list of top priorities—at that moment—to help tame the beastly number of items competing for attention. It had one scroll and could be set to focus on a single topic or all topics, but only when the topics rose to a specific level of importance. As a result, staff knew where to look and what the (distinct) sound of an alert sounded like.

Quick, Get a Heat Map...

Eye gaze heat maps show us where our users' eyes are looking on an interface. We can get a representation of the total time people spend looking at different parts of the screen; areas they focus on for longer appear to be "hotter" in other locations (Figure 9-4).

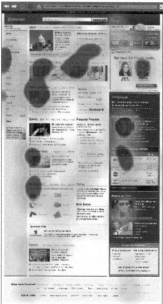

FIGURE 9-4

Heat maps

CASE STUDY: WEBSITE HIERARCHY

Challenge: In the case of the site pictured in Figure 9-4 (Comcast.net, the precursor to Xfinity) in the original version (on the left), consumers were overwhelmingly looking at one area in the upper-left corner, but not further down the page, nor at the right side of the page. We knew this both from eye tracking and the fact that the partner links further down the page weren't getting clicks (which the partners were not happy about). The problem was the visual contrast. The upper-left corner of the old page was visually much darker than the rest of the page, and more interesting (videos, images)—so much so that it was overwhelming people's visual attention systems.

Recommendation: We redesigned the page to make sure that the natural visual flow included not only the headlines, but the other information down below. We gave more visual prominence to the neglected sections of the page through balancing features like visual contrast, size of pictures, colors, fonts, and white space. We were able to draw people down visually to engage "below the fold." This made a huge difference in where people looked on the page, making end users, Comcast, and its paid advertising partners much happier.

This case study shows how helpful tools like eye tracking and heat maps can be. But I want to counter the misperception that these tools on their own are enough for you to make meaningful adjustments to your product. Similar to the survey results and usability testing that I mentioned in the last chapter, heat maps can provide you with a lot of the *what*, but not the *why* behind a person's vision and attention. The results from heat maps do not tell you what problem users are trying to solve.

To get at that, we need to...

Go with the Flow

We're trying to satisfy customers' needs as they arise, so we want to know at each stage in problem solving what our users are looking for, what they're expecting to find, and what they're hoping to get as a result. Then we can match the flow with what they're expecting to find at each stage of the process.

While observing someone interacting with a site, I'll often ask them questions like "What problem are you trying to solve?" and "What are you seeing right now?" This helps me see what's most interesting to them, at this moment, and understand their goals.

There are many unspoken strategies and expectations that users employ, which is why we can only learn through observing users in their natural flow. These insights, in turn, help us with our visual design, layout, and information architecture by clarifying what the steps are, how they should be represented, where they should be in space, etc.

CASE STUDY: AUCTION WEBSITE

Challenge: Here's an example of some of those unspoken expectations that we might observe during contextual interviews. In testing the target audience for a government auction site, I heard the feedback "Why doesn't this work like eBay?" Even though this site was even larger than eBay, our audience were much more familiar with eBay, and brought their experience and related expectations regarding how eBay worked to their interactions with this new interface.

Eye tracking confirmed users' expectations and confusion: they were staring at a blank space beneath an item's picture and expecting a "bid" button to appear, since that's where the "bid" button appears on eBay items. Even though the "bid" button was in fact present in another place, users didn't see it because they expected it to be in the same location as on eBay.

Recommendation: This was one case where I had to encourage my client not to "think different," but rather admit that other systems like eBay have cemented users' expectations about where things should be in space. We switched the placement of the button (and a few other aspects of the site architecture) to match people's expectations, which immediately improved performance. We knew *where* they were looking for this particular feature, and we knew they didn't find it in that location. This wasn't because of language or the visual design, but because of their experience with other similar sites and associated expectations.

Real-World Examples

I don't know if you've had a chance to put my Post-it note categorization method into practice yet, but I'd like to share some examples of the findings I noted in the previous chapter from clients interacting with both a video-streaming website and an e-commerce website (Figures 9-5 through 9-8). These will give you a sense of what we're looking for when subdividing data according to the Six Minds; in this case, focusing on vision and attention. Remember, there's often overlap, but I'm most concerned with the biggest problem underlying each comment:

"Can't find the 'save for later' feature."

In this case, the user was looking for a certain feature on the screen and couldn't find it, implying a visual challenge. In processing this feedback, we want to consider if a "save for later" feature was indeed present, and if so, why this participant was unable to find it. If the feature was there but was named something else (e.g., "keep" or "store for later"), this would be a language issue. Before making any changes, we would want to know if other participants had a similar issue. However, if the feature was indeed present, yet the customer's attention was not attracted to it, then *yes indeed*, this would be a vision/attention issue. Just note that some comments related to "finding" in a visual scene are not necessarily visual issues (e.g., they might signify language or other issues).

Can't find the "Save for Later" feature.

FIGURE 9-5
Research observation: participant unable to find a "Save" feature

Warning

In reviewing your findings, you're going to see a lot of comments about "seeing," "finding," "noticing," etc. Such words might suggest vision, but beware of placing such findings in the "Vision" category automatically! In reviewing each finding, ask yourself if it implies an expectation of how things should be (Memory), how to navigate through space (Wayfinding), or how familiar the user is with the product (Language) before taking the term literally and putting that observation in the Vision category.

"Homepage busy and intimidating. 'This is a lot!'"

This sounds like it is related to vision and attention. We should review the organization of this page and its information density.

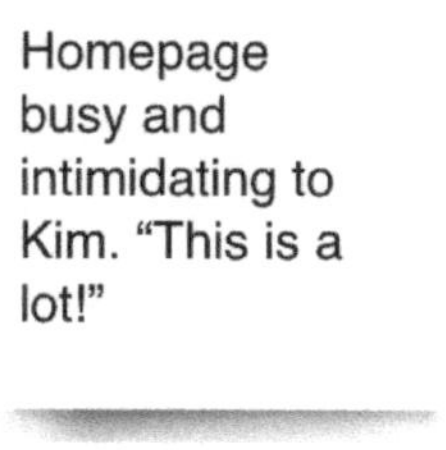

FIGURE 9-6

Homepage busy and intimidating to Kim. "This is a lot!"

Research observation: participant notes the homepage's visual complexity

"Viewed results but didn't see 'La La Land'"

It sounds like the user missed something on the page. In this example, we know the movie *La La Land* appeared in the search results, but it didn't pop out to the user. For some reason, the visual features of the search results (think back to the examples of visual "popout" that we looked at in Chapter 2, like shape, size, orientation, etc.) weren't as captivating as they should have been. Perhaps there wasn't enough visual contrast between the different search results, or there wasn't an image to draw the user's attention. Or maybe the page was just too distracting. You can take this type of feedback straight back to your visual designer. The video of this situation might be especially valuable to indicate what improvements might be made.

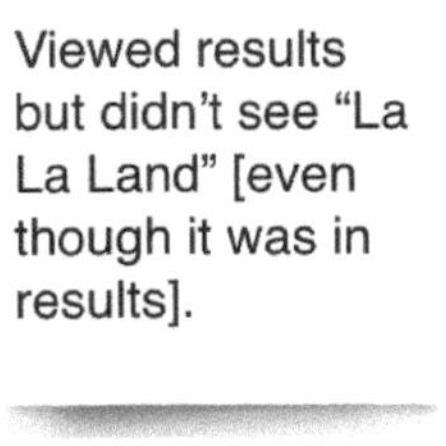

FIGURE 9-7

Viewed results but didn't see "La La Land" [even though it was in results].

Research observation: participant failed to notice results on the page

"Didn't notice 'Return to results' link. Looking for a 'back' button."

Here's a great example of the types of nuance we need to pay attention to. When you read "didn't notice," you might automatically assume this is about vision. But don't be fooled—there could be a language component at issue as well. To determine which it is (vision or language), you would need to do some sleuthing using your observational data and/or eye tracking to see where the user was looking at this moment. If they were staring right at the "return to results" link and it was still not working for them, then you know it's a language problem—those words didn't trigger the semantic content they were looking for (i.e., wrong words).

Once you've reviewed all of your customers' feedback and distilled the major problem to address, you can provide this to the visual design team with quite specific input and recommendations for improvement.

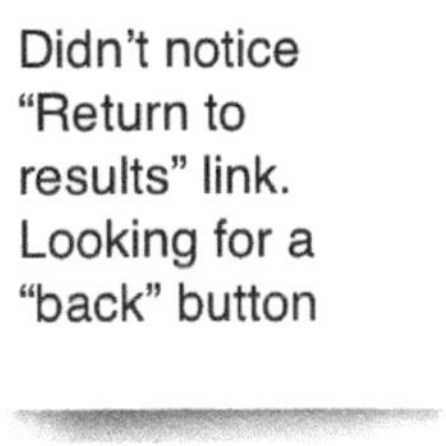

FIGURE 9-8

Didn't notice "Return to results" link. Looking for a "back" button

Research observation: participant failed to see button, looking for a different term

Concrete Recommendations

- Sit perpendicular to the participant—watch where they are looking on screens and interfaces for the next steps.

- Determine *what* they are looking for and *why* (what is most relevant to this participant at this moment, and what do they think will be good about finding that?).

- What are the assumptions they have about the system that justify those anticipations (e.g., "I'm looking for the menu because I want to make this word bold, but I don't see any menu up here; I just see all the words")?

- What else does their pattern of interaction suggest about their assumptions and unspoken strategies concerning this system?

- Build a mental model of the participant's thought process, from their perspective, watching their eye movements and behavior.

[10]

Language: Did They Just Say That?

IN THIS CHAPTER, I'LL give you recommendations on how to record and analyze interviews, paying special attention to the words people utter, their sentence construction, and what this tells us about their level of sophistication in a subject area.

Remember, when it comes to language (Figure 10-1), we're considering these questions:

- What are the words your customers use the most?

- What meaning are they associating with those words?

- How sophisticated are the words they are using? What level of expertise in the subject matter does that imply?

- Are they using the same lexicon that we (as the internal product designers) use? Are we using any jargon that they might find difficult to understand?

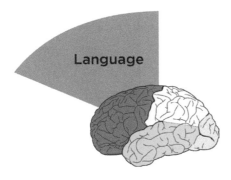

FIGURE 10-1
Language and
language processing

Recording Interviews

As I've mentioned, when conducting contextual inquiries, I recommend recording interviews. This does not have to be fancy. Sure, wireless mics can come in handy, but in truth, a two hundred dollar camcorder can actually give you quite decent audio. It's compact and unobtrusive, making it a great way to record interviews, both audio and video. Try to keep your setup as simple as possible so it's as unobtrusive to the participant as possible and minimally affects their performance. With a little creativity, you can play back the video now in a number of tools (e.g., a Zoom conference) and get a transcription, too.

Prepping Raw Data: But, But, But...

After recording your interviews, if you analyze those transcripts for word usage and frequency, you're going to find that, unedited, the top words that people use are "but," "and," "or," and other completely irrelevant words. Obviously what we care about is the words people use for representing ideas. So strip out all the conjunctions and other small words to get at the words that are most relevant to your product or service. Then examine how commonly used *those* words are.

Often, word usage differs by group, age, life status, etc.; we want to know those differences. We also want to get, through the words people are using, a sense of how much they really understand the issue at hand. There are new tools popping up all the time to measure word frequency in a passage of text. I'd recommend you Google "word frequency analysis" to find the latest.

Reading Between the Lines: Sophistication

The language we use as product and service designers can make a customer either trust or mistrust us. As customers, we are often surprised by the words that a product or service uses. Thankfully, the reverse is also true; when we get the language right and synchronize our words with those of our customers, they are more confident in what we're offering.

When we read between the lines of what someone is saying, we can "hear" their understanding of the subject matter, and this tells us about their level of sophistication. Ultimately, this leads to the right level of discussion to be having with this customer about the subject matter.

This holds for digital security or cryptography, or scrapbooking, or French cuisine. All of us have expertise in one thing or another, and use language that's commensurate with that expertise. I'm a DSLR camera fan, and love to talk about "F stops" and "anaphoric lenses" and "ND filters," none of which may mean anything to you. I'm sure you have expertise in something I don't, and I have much to learn from you.

As product and service designers, what we really want to know is, what is our typical customer's understanding of the subject matter? Then we can level-set the way we're talking with them about the problem they're trying to solve.

In the tax world, for example, we have tax professionals who know that Section 368 of "the code" (the US Internal Revenue Service tax code) is all about corporate reorganizations and acquisitions, and might know how a Revenue Procedure (or "Rev Proc") from 1972 helps to "moderate the cost-basis" for a certain tax computation. These individuals are often shocked that other humans aren't as passionate about the tax code, and are horrified that some people just want TurboTax to tell them how to file without revealing the inner workings of the tax system in all its complexities. TurboTax speaks to nonexperts in terms they can understand—e.g., What was your income this year? Do you have a farm? Did you move?

The bottom-line message here is, learn what your customers are saying, what that implies about their expertise in the subject matter, and meet your customers where they are, using terms they can understand.

CASE STUDY: MEDICAL TERMS

Challenge: You may have heard of MedlinePlus, which is a part of NIH.gov. The website, depicted in Figure 10-2, provides an excellent and comprehensive list of different medical issues. The challenge we found for users of the site was that MedlinePlus listed medical situations by their formal names, like "TIA" or "transient ischemic attack," which is the accurate name for what most people would call a "ministroke." If NIH had only listed TIA, the average user of the site would likely be unable to find what they were looking for in a list of search results.

FIGURE 10-2
MedlinePlus

Recommendation: We advised the NIH to have its search function work both for formal medical titles and more common vernacular. We knew that both sets of terms should be prominently presented, because if someone was looking for "mini-stroke" and didn't see it immediately, they would probably feel like they got the wrong result. A lot of times, experts internal to a company (and doctors at NIH, and tax accountants at a law firm) will struggle with including more colloquial language because it may not be strictly accurate, but I would argue that as designers, we should lean more toward accommodating the novices than the experts if we must choose between the two options. Or, if you can, follow the style of Cancer.gov that I mentioned in Chapter 5, where for each medical condition users have the choice of viewing either the health professional version or the patient version.

Real-World Examples

Looking again at our sticky notes, Figures 10-3–10-5 show the ones we should place in the Language column:

"Couldn't find the shopping cart. Eventually figured out the 'Shopping bag' is the cart."

If we know that a "shopping cart" feature is present on our site, there are two possible reasons why the participant was unable to find it: (a) there was a visual issue that prevented them from actually seeing the feature, or (b) they were staring at the correct feature, yet did not understand that what they were looking at was what they were looking for because they were expecting to see a different term (i.e., "shopping bag" versus "shopping cart"). To discern which your site should have, you'll want to consult your notes or video footage to see where the user was actually looking at this moment. In this case, my notes indicated that the user eventually figured out the "shopping bag" was the "cart," suggesting that this was indeed a language issue.

This one is a great example of how the same word in English can mean different things in America, versus Canada, versus the UK, etc. Many of us in America say "shopping cart" with visions of Costco and SUV-sized carts in our heads, but in many parts of the world where public transit is the norm, shopping bags prevail. With this type of finding, we would want to know if other participants had a similar issue to determine if we should change the terminology.

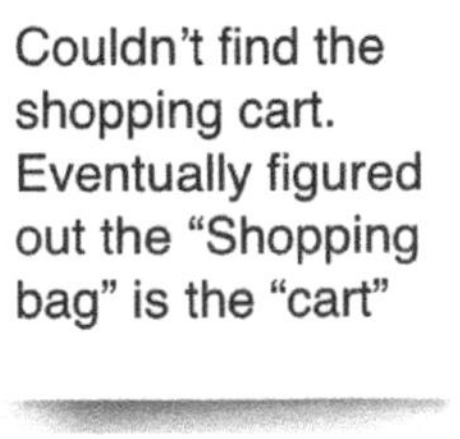

FIGURE 10-3

Research observation: participant using different terms

Couldn't find the shopping cart. Eventually figured out the "Shopping bag" is the "cart"

"Searched for 'Eames Midcentury Lounge Chair' when asked to search for a chair."

Here, I would argue it's highly unlikely for the average shopper to know that there is something called an Eames chair, and that it's a lounge chair, and that it's a mid-century lounge chair. These search terms suggest to me that this person is extremely knowledgeable about mid-century modern furniture, which demonstrates the high level of expertise of this particular shopper and the type of language we might need to use to reach them. If this is a trend, we need to let content specialists know to accommodate this level of sophistication.

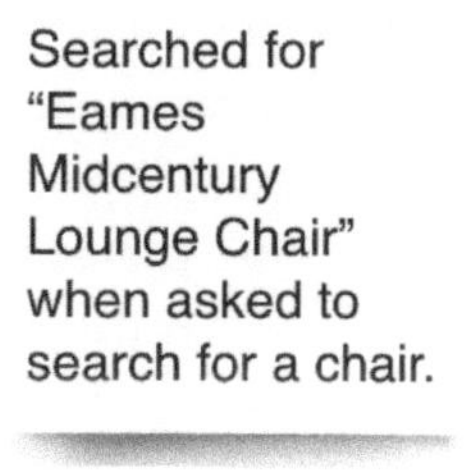

FIGURE 10-4

Searched for
"Eames
Midcentury
Lounge Chair"
when asked to
search for a chair.

Research observation:
search terms
suggest considerable
knowledge of this
domain

Warning

Someone interpreting these findings too literally might consider "searching" a
visual function (i.e., scanning a page up and down to find what you're looking
for), whereas this instance of "searching" implies typing something into a
search engine. As always, if you're unsure about a comment or finding that's
out of context, go back to your notes, video footage, or eye tracking, and see
if the user is literally searching all over the page for a bike, or typing something
into a search engine. When you take notes, make sure to be as clear as possible
with words like "search" that can be interpreted in multiple ways.

"Wants to filter results by 'Film noir.'"

> This data point speaks to the person's expertise in the field and
> the sophistication of the language and background understanding
> they have regarding the film industry. There's also an element of
> memory here, as it points to the user's mental model of having
> results organized by genre, perhaps the way a similar site orga-
> nizes its results.

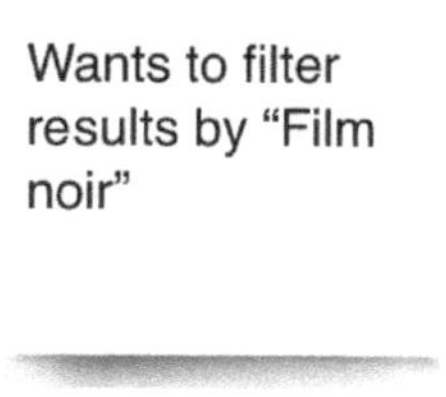

FIGURE 10-5

Research observation:
search term suggests
deeper knowledge of
film industry

Wants to filter
results by "Film
noir"

Through this exercise, I've given you a small taste of the breadth of the types of language responses we're looking for. They range from mis-named buttons to cultural notions of word meanings to nomenclature associated with an interaction/navigation to level of sophistication.

Contrast your customers' word usage with your web or app's language, and consider how this might impact your product or service design. Due to the range of expertise that your customers' language demon-strated, you might conclude we need the design to better reflect the needs of both novices and experts on the site.

CASE STUDY: INSTITUTE OF MUSEUM AND LIBRARY SERVICES

Challenge: The example in Figure 10-6 demonstrates the importance of appropriate names for links, not just the content. This govern-ment-funded agency, which you probably haven't heard of, does amaz-ing work supporting libraries and museums across the U.S. If you look at the organization of the website navigation, it's pretty typical: About Us, Grants, Publications, Research & Evaluation, and... Issues. When we tested this with users, what caught their eye was the Issues tab. All of our participants assumed "Issues" meant things that were going wrong at the Institute. This couldn't be further from the truth; the Issues area actually represented the Institute's areas of top focus and included discussion of topics relevant to museums and libraries across America (preservation, digitization, accessibility, etc.).

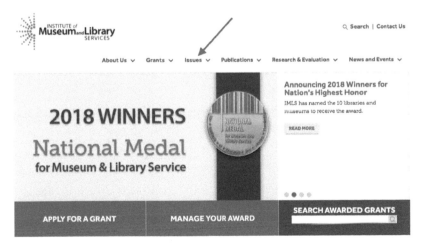

FIGURE 10-6
The Institute of Museum and Library Services website

Recommendation: The key point here is that we needed to consider matching not only the language in general to customer expectations, but also the navigational terminology. Moving forward, the Institute will move the "Issues" content to another location with a new name that better conveys the underlying content.

Concrete Recommendations

- Record all interview audio and transcribe it using an automated tool.

- Measure the frequency of word use and the level of sophistication of the vocabulary (e.g., "hurt brain" versus "left medio-parietal subdermal hematoma") to get an indication of the sophistication of the customers' understanding of an issue.

- Study and pay attention to word order and words used, especially when building AI systems (in order to get the appropriate training in and ensure certain syntactic patterns will be correctly processed).

[11]

Wayfinding: How Do You Get There?

Now LET'S TURN TO findings that are related to wayfinding (Figure 11-1). As a reminder of what we discussed in Chapter 2, wayfinding is all about where people think they are in space, what they think they can do to interact and move around, and the challenges they might have there. We want to understand people's perception of space—in our case, virtual space—and how they can interact in that virtual world.

Remember the story about the ant in the desert? It was all about how it thought it could get home based on its understanding of how the world works (which lacked accounting for being picked up and moved). Similarly, we want to observe the navigation and wayfinding behavior of our customers and identify any issues they are having while interacting with our products and services.

With wayfinding in mind, we are seeking to answer these questions:

- Where do customers *think* they are?

- How do they *think* they can get from Place A to Place B?

- What do they *think* will happen next?

- What are their expectations, and what are those expectations based on?

- How do their expectations differ from how this interface actually works?

- What interaction design challenges did they encounter as a result of their assumptions?

In this chapter, we'll look at how customers "fill in the gaps" with their best guesses of what a typical interaction might be like, and what comes next. Especially with regard to service designs and flows, we need to know our customers' expectations and anticipate their next steps so we can build trust and match those expectations.

FIGURE 11-1
Wayfinding, generally thought to harness the parietal lobes

Wayfinding

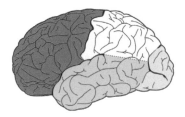

Where Do Users Think They Are?

Let's start with the most elemental part of wayfinding: where users think they actually are in space. Often with product design, we're talking about virtual space, but even in virtual space, it's helpful to consider our users' concept of physical space.

CASE STUDY: SHOPPING MALL

Challenge: You need to know where you are in order to determine if you've reached your destination, or, if not, how you will get there. Returning to the picture of that mall near my house, you can see that everything is uniform: the chairs, the ceiling, the layout (Figure 11-2). You can't even see many store names. This setup gives you very few clues about where you are and where you're going (physically and philosophically, especially when you've spent as much time as I have trying to find my way out of shopping malls!). It's a little bit like the Snapchat problem we looked at in Chapter 3, but in physical space: there's no way to figure out where you are, no unique cues.

FIGURE 11-2
What cues are you getting about your location in this mall?

Recommendation: I've never talked with our mall's design team, but if I did, I would probably encourage them to add some unique cues, such as different colored chairs on different wings, or to remove some of the poles that block customers from seeing the stores ahead of them. All we need are a few cues that can remind us where we are and which way to go. The same goes for virtual design: Do you have virtual signposts or cues in place so your customers can tell where they are in the virtual space? Are the entrances, exits, and other key junctions clearly marked?

How Do They Think They Can Get from Place A to Place B?

Just by observing your users in the context of interacting with your product, you'll notice the tendencies, workarounds, and "tricks" they use to navigate. Often, this happens in ways you never expected when you created the system in the first place.

CASE STUDY: SEARCH TERMS

Challenge: Something I find remarkable is how frequently users of expert tools and databases actually start out by Googling words or phrases (terms of art) they think might come in handy while using those high-end tools. In observing a group of tax professionals, I realized that they thought they needed a specific term of art (i.e., a certain tax code term) to find the right page of tax code in a tool we were using, and that it couldn't be found by browsing the system. Instead of navigating to the tax code, they Googled the name of the tax law to identify the term used by specialists, returned to the tool, and typed that term into the search bar. As designers, we know it's because they were having trouble navigating the expert tool in the first place that they found other ways around that problem.

Recommendation: In designing our products or services, we need to make sure we take into account not only our product, but the constellation of other "helpers" and tools—search engines are just one example—that our end users are employing in conjunction with our product. We need to consider all of these to fully understand the big picture of how they believe they can go from Point A to Point B.

What Are Those Expectations Based On?

As you'll notice as you embark on your own contextual inquiry, there is a lot of overlap between wayfinding, language, and memory; after all, any time someone interacts with your product or service, they come to it with base assumptions from memory.

Let me try to draw a line between wayfinding and memory. When talking about memory, I'm talking about a big-picture expectation of how an experience works (e.g., dining out at a nice restaurant, or going to a car wash). With wayfinding, or interaction design, I'm talking about expectations relating to moving around in space (real or virtual).

Here's an example of the nuanced differences between the two. In some newer elevators, you have to type the floor you're headed to on a central screen outside the elevator bank, which indicates which elevator you should take to get there. There are no floor buttons inside the elevators. This violates many people's traditional notions of how to get from the lobby to a certain floor using an elevator. But because this relates to moving around in space, even though it taps into stored memories and schemas, I'd argue this is an example of wayfinding. In this case, the memory being summoned up is about an interaction design (i.e., getting from the lobby to the fifth floor), as opposed to an entire frame of reference.

Real-World Examples

Getting back to our sticky notes, Figures 11-3–11-7 show the ones we would categorize as findings related to wayfinding:

"Expected the search to provide type-ahead choices."
> This isn't analogous to an ant moving around in space, but I do think that it relates to interaction design. It does use the word "expected," implying memory, but I think the bigger thing is that it's about how to get from Point A (i.e., the search function) to B (i.e., the relevant search results).

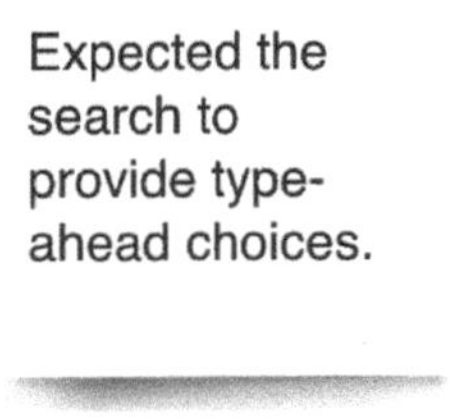

FIGURE 11-3
Research observation: search interaction expectations

Expected the search to provide type-ahead choices.

"Expected that clicking on a book cover would reveal its table of contents."
> That's an expectation about interaction design. This person has specific expectations of what will happen when they click on a book cover. This may not be the way most electronic books work right now, but it's good to know that's what the user was expecting.

FIGURE 11-4
Research observation: ecommerce interaction expectation

Expected that clicking on a book cover would reveal its table of contents.

"Expecting to be able to 'swipe like her phone' to browse."

Here's an example of wayfinding we are seeing more and more as we work with "digital natives." Like most of us, this person uses her phone for just about everything. As such, she expected to swipe, like on a phone, to browse. This sort of "swipe, swipe, swipe" expectation is increasingly becoming a standard, and something we need to take into account as designers. You could argue there's a memory/frame of reference component, but I would counter that the memory in question is *about* an interactive design and how to move around in this virtual space.

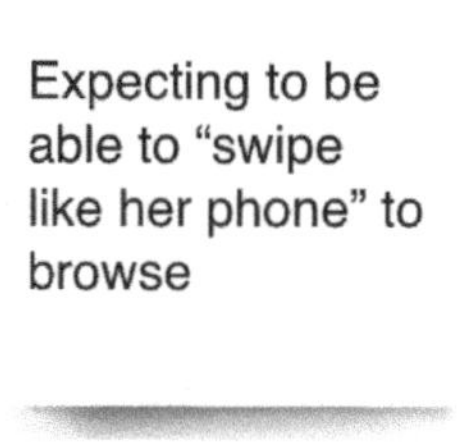

Expecting to be able to "swipe like her phone" to browse

FIGURE 11-5
Research observation: phone interaction expected on other surfaces

"Frustrated that voice commands don't work with this app."

This is a fair point about interaction design; the user would like to use voice interactions in addition to, say, clicking somewhere or shaking their phone and expecting something to happen. This is a good example of how wayfinding is about more than just physical actions in space. You might argue there is a language component here, but we're not really sure this user had the expectation of being able to use voice commands; just that they would have liked it. We could get more data to know if a memory of another tool was responsible for this frustration.

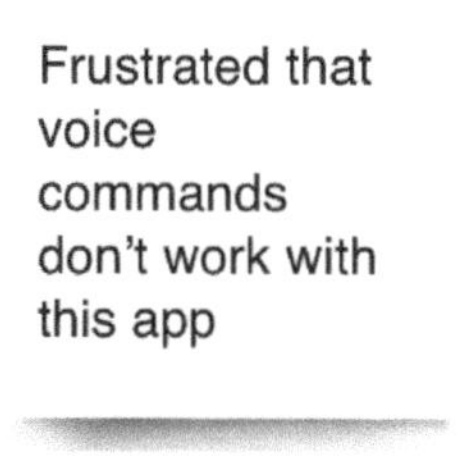

Frustrated that voice commands don't work with this app

FIGURE 11-6
Research observation: participant wanted voice-based interaction

"Expects that clicking on the movie starts a preview, not actual movie."

This person had expectations of how to start a movie preview on a Roku or Netflix-type interface. In this particular case, it sounds like you get either a brief description or the whole movie, and nothing in between, which violates the user's wayfinding expectations. If a preview option were there but the user missed it for some reason, we would recategorize this as a visual issue.

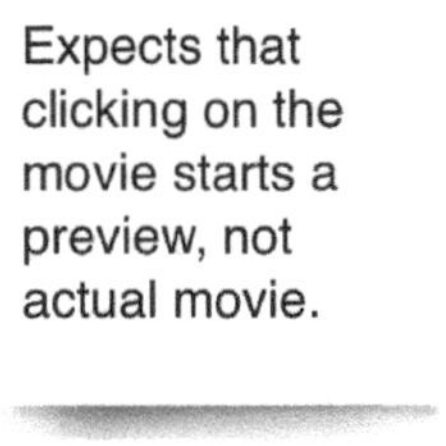

FIGURE 11-7
Research observation: expectations based on past experience

Expects that clicking on the movie starts a preview, not actual movie.

Case Study: Distracted Movie-Watching

Challenge: Since we're on the topic of phones, I thought I'd mention one study where I observed participants looking at their phone, and a TV, and also how they navigated from, say, the Roku to other channels like Hulu, Starz, ESPN, etc. In this case, we were interested in how participants (who were wearing eye tracking glasses, as seen in Figure 11-8) thought they could go from one place to another within the interface. (Are they going to talk to the voice-activated remote? Are they going to click on something? Are they going to swipe? Is there something else they're going to do?)

Recommendation: Two things came out loud and clear here. First, the "flat" design style that is so typical is not great. It is often difficult for users to know which element on the screen is currently selected, so they have real trouble knowing "where they are" in the interface. Second, the Roku was head and shoulders above the rest. Why? Because of one element on the remote: a back button! No matter where they were in the interface or on a channel, the back button worked exactly the same way. A great example of matching customer predictions about site navigation!

FIGURE 11-8
Using a head-mounted eye tracker to study attention on a TV-based interface

Concrete Recommendations

- Ask users before anything happens with a system how they think it will work and why. Learn as much as you can about user expectations.

- Ask these questions throughout the contextual interview: What will happen next? What will you have to do? What will happen if you make a mistake? How will you know it worked?

- When a step is taken, moderators can ask (often knowing the answer but not the explanation): Is that what you expected would happen? Why/why not? What should have happened? Did that surprise you?

[12]

Memory: Expectations and Filling in Gaps

IN THIS CHAPTER, WE'RE going to consider the semantic associations our customers have. By this, I mean not just words and their meanings, but also their biases and expectations (Figure 12-1).

Some of the questions we'll ask include:

- What are the frames of reference our audience is using?

- What were they expecting to find?

- How were they expecting this whole system to work?

- What are the stereotypes, mental models, or schemas influencing those expectations?

- How do the customers' stereotypes differ from our expert schemas or stereotypes?

- What changes can we make to ensure that we're meeting customers' expectations?

FIGURE 12-1
Memory consolidation is generally thought to occur in lower brain regions.

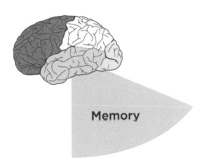

Memory

Meanings in the Mind

Let's go back to the idea of stereotypes, which I mentioned in Chapter 3. These aren't necessarily negative, as the stereotypical interpretation of "stereotype" would have you believe—as we discussed earlier, we have stereotypes for everything from what a site or tool should look like to how we think certain experiences are going to work.

Take the experience of eating at a McDonald's, for example. When I ask you what you expect out of this experience, chances are you're not expecting white tablecloths or a *maître d'*. You're expecting to line up, place your order, and wait near the counter to pick up your meal. At a more modern McDonald's, you might also expect a touchscreen ordering system. I know someone who recently went to a test McDonald's and was shocked that they got a table number, sat down at their table, and had their meal brought to them. That experience broke this person's stereotype of how a quick-serve restaurant works.

For another example of a stereotype, think about buying a charging cable for your phone on an e-commerce site, versus buying a new car online. For the former, you're probably expecting to have to select the item, indicate where to ship it, enter your payment information, confirm the purchase, and receive the package a few days later. In contrast, with a car-buying site, you might select the car online, but you're not expecting to buy it online right then. You're probably expecting that the dealer will ask for your contact information to set up a time for you to come in and see the car. Buying the car will involve sitting down with people in the dealership's financial department. These are two very different expectations for how the interaction of purchasing something will go.

CASE STUDY: PRODUCING THE PRODUCT VERSUS MANAGING THE BUSINESS

Challenge: For one project with various small business owners we noticed that, broadly speaking, our audience fell into one of two categories:

1. Passionate producers

   They loved the craftwork of producing their product, but weren't as concerned with making money. They were all about making the most beautiful objects they could and having people love their work. They loved building relationships with their customers.

2. Business managers

> They didn't care as much about what they were selling and its craftsmanship; they were looking much more at running the business and making it efficient. They weren't as client-facing.

Outcome: We realized there isn't necessarily one set of expectations from small business owners. Rather, there are two hugely different patterns here, with unique expertise and expectations. One group loved working up forecasts in spreadsheets, whereas the other group wanted nothing to do with that. One group was great at customer engagement; the other group preferred to work behind the scenes. Each group had different expertise, and therefore needed products, services, and language that catered to its respective strengths. This story goes to show that identifying the different types of audiences you have can really help to influence the design of your product. More on audience segmentation coming in Part III!

Putting It All Together

We want to consider all facets of people's preprogrammed expectations: how something should look, how that thing should work, how to get to the next step in the process, and generally how customers think the whole system will work. As you work to decode your customers' expectations about the product or service experience, I want you to look out for how different novices can be from experts. As we've been discussing with language, it's crucial that we meet users at their level.

CASE STUDY: TAX CODE

Challenge: For one client, we observed accountants and lawyers doing tax research. Specifically, we looked at how they were searching for particular tax code information. While the information was historically organized by the type of publication (e.g., journal, book), the end users' expectations and mental organization revolved around very different dimensions (US taxes versus. international taxes, estate tax versus corporate tax, guides versus tax law, etc.). The interaction models that were available to them at the time just weren't matching the multidimensional representation they were seeking.

Recommendation: In creating a tool that would be most helpful to this group, the designers needed to revise their model to be more in line with users' thinking and provide them with filters that matched their mental model (Figure 12-2).

Your audience is thinking...

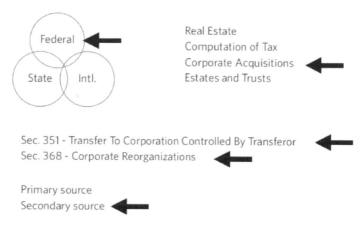

FIGURE 12-2

Tax experts have a multidimensional organization of tax law information

Real-World Examples

Going back to our sticky notes, let's think about memory and which findings relate to that lens of our mind (Figures 12-3, 12-4, and 12-6):

"I want this to know me like Stitch Fix."

This one involves a frame of reference. (For context, Stitch Fix is a fashion service that sends you clothes monthly, similar to Trunk Club. After you give it a general sense of your style, the service uses experts and computer-generated choices to create your shipment, which you try on at home, paying for the items you like and sending back the others.) There's definitely some emotion going on here, suggesting that the user wants to feel known when using our tool, and is perhaps expecting a certain type of top-flight, professional customer experience. But I think the main point of this finding is that it suggests the user's overall frame of reference in approaching our tool. Knowing the user's expected model of interaction is helpful for us to discern what this person is actually looking for in our site.

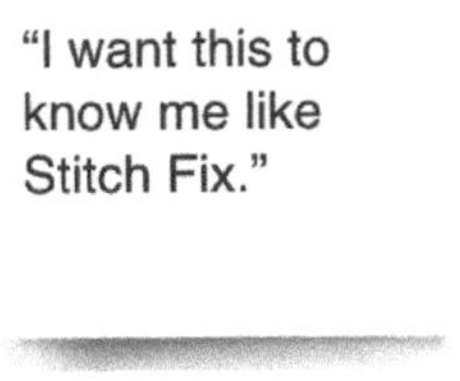

FIGURE 12-3

Research observation:
wide range of
expectations based
on experience with
another site

"I want this to
know me like
Stitch Fix."

"Can't figure out how to get to the 'checkout counter.'"

Here, it sounds like the user is thinking of somewhere like Macy's, where you go to a physical checkout counter. You might argue that this is vision, since the user is searching for something but can't find it; you might argue this is language because they're looking specifically for something called the "checkout counter"; you could argue it's wayfinding because it has to do with getting somewhere. Or is it memory? None of these are wrong, and you'd want to consult your video footage and/or eye tracking data if possible. But I think the key point is that the user's perspective (and associated expectations, of a brick-and-mortar store) were way off what a website would provide, implying a memory and expectations issue.

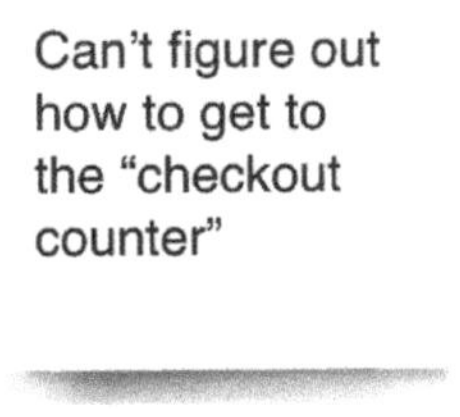

FIGURE 12-4

Research observation:
participant's
expectations are
inconsistent with
current design

Can't figure out
how to get to
the "checkout
counter"

This comment makes me think of a fun tool you all should check out called the *Wayback Machine*. This internet archive allows you to go back and see old iterations of websites. This idea of a checkout counter comment reminds me of an early version of the Southwest Airlines website (Figure 12-5).

Subscribe to Southwest Airlines' Click 'n Save℠ Updates for the latest fare specials!

Southwest Airlines Home Gate
The Home of Southwest Airlines on the World Wide Web

Updated February 23, 1999:

FIGURE 12-5

Southwest.com in its first iteration

As you can see, Southwest was trying to be very compatible in its early web designs with the physical nature of a real checkout counter. What they ended up with was this overly physical representation of how things work at a checkout counter, with a weigh scale, newspapers, etc. All of these features were incredibly concrete representations of how a checkout counter works. Even though digital interfaces have abandoned this type of literal representation (and most of these physical interactions have gone away, too), it's important to keep prior patterns of behavior in mind when designing for older audiences whose expectations may be more in line with the checkout counter days of old.

[SIDE NOTE]

Because the "checkout counter" comment is so unique (and was only mentioned by one participant), we might not address this particular item in our design work. In reviewing all your feedback together, you'll come across instances like these where you chalk it up to something that's unique to this individual, and not a pattern you're seeing for all of your participants. When taken in context with the totality of this person's comments, you may also realize that this was a novice shopper (more on that in Part III, where we'll look at audience segmentation).

"Expects to see 'Rotten Tomato' ratings for movies."

I think this one also points to an expectation of how ratings work on other sites, and how that expectation influences the user's experience with our product. This is also another example of where too literal a reading of the findings may mislead you. When you read "expects to see," beware of assuming the word "see" indicates vision. In this case, I think the stronger point is that the user has a memory or expectation about what should be on the page. You could argue that the expectation is linked to their wanting to make a decision, so I would say this one could fall under either memory or decision making.

FIGURE 12-6

Expects to see "Rotten Tomato" ratings for movies.

Research observation: participants referencing past experience want an equivalent, not the same literal content

What You Might Discover

In our exercise, we looked at users' expectations about other tools, products, and companies; how our users are used to interacting with them; and how users carry those expectations over into how they expect to work with our product, or the level of customer service they're expecting from us. These are the types of things we're usually looking for with memory. We want to look out for moments of surprise that reveal our audience's representation, or the memories that are driving them. We also talked a little about language and level of sophistication, as these can indicate our users' expectations.

We want to understand our users' mental models and activate the right ones so that our product is intuitive, requiring minimal explanation of how it works. When we activate the right models, we can let our end audience engage those conceptual schemas they have from other situations to do what they need to do. This helps build more trust in our product or service.

CASE STUDY: TIMELINE OF A RESEARCHER'S STORY

Challenge: For one client, we considered the equivalent of LinkedIn or Facebook for a professor, researcher, graduate student, or recent PhD looking for a job. As you may know, Facebook can indicate your marital status, where you went to school, where you grew up, even the movies you like. In the case of academics, you might want to list mentors, what you've published, if you've partnered in a lab, and so on. We realized there were a lot of pieces when it came to representing the life and work of a researcher.

Recommendation: Doing contextual inquiry revealed what a junior professor would like to see about a potential graduate student, and how a department head might review the results of someone applying for a job in a very different way. We learned a lot about users' expectations regarding categories of information they wanted to see and how it should be organized, and how that differed from a typical resume or curriculum vitae (Figure 12-7).

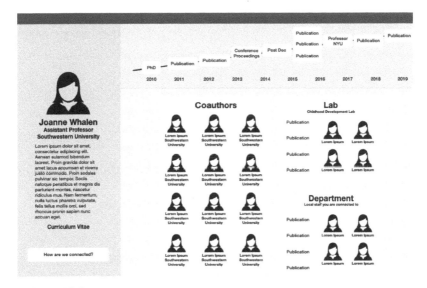

FIGURE 12-7
Hypothetical web page for an academic profile

Concrete Recommendations

- Ask customers about the underlying source of their expectations: What are you basing your expectations on? What else have you used that works like this?

- Build not just an audience persona (more on this in Part III), but a set of assumptions that persona holds regarding word usage, how things should work according to their expectations, and how they are framing the problem.

- Document visual attention biases and the words/actions users look for, word usage and the meanings associated with those words, the syntax of their sentence construction, the answers they are expecting from the system, and the flow that users expect to have.

- All of these together can provide powerful suggestions for how best to match the system to end users' needs.

[13]

Decision Making: Following the Breadcrumbs

When it comes to decision making, we're trying to figure out what problems customers are really trying to solve and the decisions they have to make along the way (Figure 13-1). What are they trying to accomplish, and what information do they need to make a decision at this moment in time?

With decision making, we're asking questions like:

- What is the user trying to accomplish?
- What does their overall decision-making process look like?
- What facts do they need to make their decision and solve their problem?
- What do they need at each stage of problem solving?
- When do they seem to be overwhelmed and "satisfice"?
- What middle-of-the-road, "sensible" option do they default to?

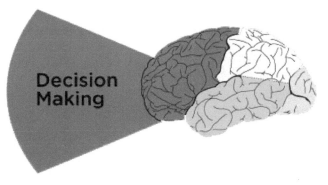

FIGURE 13-1
Decision making

What Am I Doing? Goals and Journeys

We want to focus on all the subgoals customers need to accomplish to get from their initial state to their final goal.

Your customer's end goal might be to make a cake, but to get there, they are embarking on a journey with quite a few steps along the way. First, they are going to need to find a recipe, get all the ingredients, and put them together according to the recipe's instructions. Within the recipe, there are many more steps—turning on the oven, getting out the right-sized pan, sifting the flour, mixing the dry ingredients together, etc. We want to identify each of those microsteps that are involved with our products, and how we can support our end customers in making their ultimate decision or reaching their goal.

CASE STUDY: ECOMMERCE PAYMENT

Challenge: For one client, we observed a group that was trying to decide which ecommerce payment tool to use (PayPal, Stripe, etc.). Through interviews, we collected a series of questions or concerns that people had and—you guessed it—wrote them all down on sticky notes (e.g., "What if I need help?", "How does it work?" "Can this work with my particular ecommerce system right now?" "What are the security issues?") There were so many microdecisions to make and questions people wanted answered before they were willing to proceed in acquiring one of these tools.

Outcome: By capturing each of these subgoals and ordering them, you can ensure your designs support the customer journey, answering each set of questions in a timely fashion. This helps your customers to trust your product/service and make a decision—ultimately giving them a better overall experience where they feel like they're making an informed choice.

Gimme Some of That! Just-in-Time Needs

You may remember from Chapter 5 that we can be very susceptible to nonideal psychological influences when making decisions (which is why I never let myself sit in a car I'm not intending to buy). Often we get overwhelmed with choices and end up defaulting to satisficing—accepting an available option as satisfactory. That's why people were more willing to buy the $349 blender when it was displayed in between one for $199 and one for $499. Choosing the middle option seems sensible to people. We want to keep this type of classic framing problem in mind as product designers, considering what the "sensible" option may be for our customers.

CASE STUDY: TEACHER TIMELINE

Challenge: We looked at a group of teachers and what they sought at different times of the year in terms of continuing education and support from experts that could improve their teaching skills and strategies. We learned that depending on the time of year, the support these teachers wanted was drastically different (Figure 13-2).

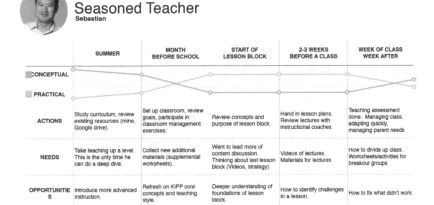

Seasoned Teacher
Sebastian

	SUMMER	MONTH BEFORE SCHOOL	START OF LESSON BLOCK	2-3 WEEKS BEFORE A CLASS	WEEK OF CLASS WEEK AFTER
CONCEPTUAL					
PRACTICAL					
ACTIONS	Study curriculum, review existing resources (mine, Google drive).	Set up classroom, review goals, participate in classroom management exercises.	Review concepts and purpose of lesson block.	Hand in lesson plans. Review lectures with instructional coaches.	Teaching assessment done. Managing class, adapting quickly, managing parent needs
NEEDS	Take teaching up a level. This is the only time he can do a deep dive.	Collect new additional materials (supplemental worksheets).	Want to lead more of content discussion. Thinking about last lesson block (Videos, strategy).	Videos of lectures. Materials for lectures.	How to divide up class. Worksheets/activities for breakout groups
OPPORTUNITIES	Introduce more advanced instruction.	Refresh on KIPP core concepts and teaching style.	Deeper understanding of foundations of lesson block.	How to identify challenges in a lesson.	How to fix what didn't work

FIGURE 13-2
Journey map representing a teacher's focus and interests throughout an academic year

Outcome: In the summer, teachers had more time to digest concepts and foundational research concerning the philosophy of education. This was the best time for them to focus on their own development as teachers and consider the *why* behind their teaching methods. Right before the school year started, however, they went from wanting conceptual development to wanting very pragmatic support. Students would soon be showing up after a three-month break, and teachers would have to manage them—in addition to the parents. During this time, they wanted highly practical information like worksheets. The micro-decisions they had to make were things like "Can I print this worksheet out right now?" "Does it have to be printed at all?" or "Can we use it on Chromebooks?" During the school year, they weren't concerned with the *why* but the *how*, often defaulting to satisficing when they became overwhelmed with too much information. Based on these findings, we were able to recommend drastically different types of content at different points of the year.

Chart Me a Course: The Decision-Making Journey

We want to know not just the overall decision our customers are making, like whether to buy a car, but also all the little decisions they have to make along the way. "Does it have cup holders?" "Will my daughter be happy riding in it?" "Can it carry my windsurfer?" "Can I put a roof rack on the top?"

Timing of information is crucial when it comes to these microdecisions. Once we've identified the questions, we want to know when the customer need to address each of them. Usually it's not all at once, but one step at a time along the journey. That's why most ecommerce sites place shipment information at the end, for example, rather than presenting you with too much information right at the start, when you're just browsing.

We also want to know what our users think they can do to interact with the system and solve their problem. In cognitive neuroscience, we talk about "operators in a problem space," which just means the levers that we think we can move in our heads to get from where we are to where we need to be. The actual problem space may be the same as or different than what your customers are envisaging, depending on how expert they are in the subject matter.

Real-World Examples

Looking once again at the sticky notes. Here are some findings that relate to decision making (Figures 13-3 through 13-7):

"Concern: 'What if the $5,000 chair I'm buying is damaged in transit?'"
> This person is basically saying they're not going to go any further with the purchase until they understand the answer to this shipping and handling question. You could argue there's some emotional content of worry or fear here, but I would say the most important aspect is that it's one of several problems to be solved along this user's decision-making journey. This will be a blocker and needs to be resolved to the customer's satisfaction before they hit the "buy" button.

FIGURE 13-3

Concern: "What if the $5,000 chair I'm buying is damaged in transit?"

Research observation: a reason customers might not hit "buy"

"Surprised that coupons can't be entered on the product page to update price."

I see this type of comment a lot in ecommerce (we'll take a look at a case study momentarily). In physical shopping interactions, we typically hand the cashier our coupons before we pay. When ecommerce sites order the interaction differently, it can throw us off and make it difficult for us to proceed without assurance that our coupon will count. Suddenly we have no interest in buying for full price!

FIGURE 13-4

Surprised that coupons can't be entered on the product page to update price.

Research observation: the importance of matching expected shopping flow

"Wants to know right away if this site will accept PayPal."

Here's another microdecision example of something the user wants to know before proceeding any further. A lot of people have a preferred or trusted method of payment, so even though we tend to put payment information toward the end, this feedback suggests that we need some indicator early on alerting the customer to the modes of payment we accept. This is another microconsideration the customer needs to tick off before they're willing to keep going.

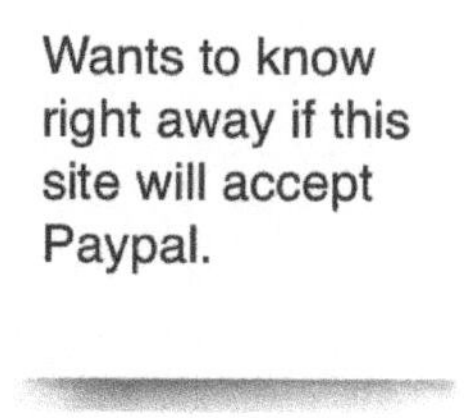

FIGURE 13-5

Wants to know right away if this site will accept Paypal.

Research observation: a microdecision to be made before purchasing

"Wants easy way to buy on laptop and send movie to TV."

This is a good example of classic problem solving, representing how the user wants to solve a problem and move around in the problem space. The problem is how to use a different tool to do the buying (of a movie) than to do the viewing. There's a bit of interaction design going on here, but I'd argue that the highest-level issue in this case is solving a problem.

FIGURE 13-6

Wants easy way to buy on laptop and send movie to TV

Research observation: a concrete problem the customer is trying to solve

"Doesn't want parents to know what she's watching."

This customer is wondering about privacy settings at this stage of her decision-making process. Maybe she wants to make sure her parents don't know she's watching horror movies before she commits to this service. Privacy considerations like what's being recorded and logged, levels of privacy, and who's receiving that data are all very legitimate concerns in our Big Data world right now.

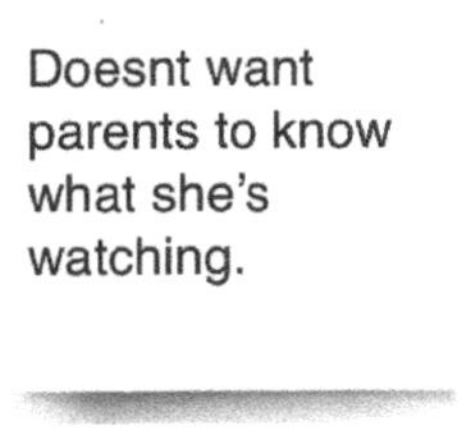

FIGURE 13-7

Doesnt want
parents to know
what she's
watching.

Research observation:
a secondary problem
that is also important
to the customer

CASE STUDY: COUPONS

Challenge: We worked with one group that offered language classes, which people could buy online. This group offered coupons, but the programmers placed the coupon code feature at the tail end of the purchasing process (it was easier to program that way). So someone buying a three hundred dollar course with a one-third-off coupon would have to first check a box indicating that they wanted to buy the course at full price, *then* put in the coupon, and then finally see the price drop to two hundred dollars. Most people would feel uncomfortable selecting the full-price course and hitting "go" without seeing confirmation that their coupon code had gone through.

Recommendation: We strongly advised the group to move the coupon code feature up in the process, so that users were checking a box that showed their applied discount. There's actually a lot of psychology behind "couponing," but I'll save that for another book.

Concrete Recommendations

- Periodically ask what users are trying to do at that moment and map out the microgoals that together suggest the expected path to the end answer (e.g., enter zip code, select movie, select date, view and select a location, view and select seats, reserve those seats).

- Build a decision-making journey map—why they are looking for information, what information they want or don't want at that step, and what they need next.

[14]

Emotion: The Unspoken Reality

WE MIGHT KNOW WHAT our customers are trying to do on a logical level (e.g., buy a car), but what are they trying to accomplish on a deeper level? What emotions do those goals or fears of failure illicit? Based on those emotions, how "Spock-like," or analytical, will a person be in their decision making?

In this chapter, we return to the last of our Six Minds: emotion (Figure 14-1). As we discuss emotion, we'll consider these questions:

- What immediate emotions are our users experiencing as they interact with our products or services?

- Which comments pertain to who this person is (i.e., their self-concept)?

- What are they trying to accomplish in life?

- What are they most afraid of having go wrong? Why?

- Who are our customers on a deeper level?

- What will make them feel accomplished?

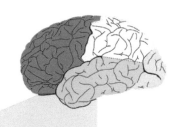

FIGURE 14-1
Emotion often forms deep within the cortex and lower systems

Emotion

Live a Little (Finding Reality, Essence)

When talking about emotion, I want you to be thinking about it on three planes:

1. Appeal

   What will draw customers in immediately? An exclusive offer? Some feature that ticks one of their microdecision-making boxes? During a customer experience, what specific events or stimuli (e.g., encountering a password prompt or using a search function) are associated with emotional reactions?

2. Enhance

   What will enhance customers' lives and provide meaningful value over the next six months, and beyond?

3. Awaken

   Over time, what will help awaken your customers' deepest goals and wishes (and support them in accomplishing those goals)? What are some of the underlying emotions they have about who they are and what they're trying to become (e.g., a good father, a millionaire, a steady worker)? What are they afraid of?

Though quite different, all of these forms of emotion are extremely important to consider in our overall experience design. We want to know what our consumers are thinking about themselves at a deep level, what might make them feel accomplished in society, and what their biggest fears are. Our challenge is then to design products for both the immediate emotional responses as well as those deep-seated goals and fears.

[SIDE NOTE]

While it may be tempting to focus on the perks of our products, we have to go back to Daniel Kahneman, who tells us that humans dislike losses more than we like gains. As such, it's extremely important to consider fear. There could be short-term fears, like not receiving a product in the mail on time, but there are also longer-term fears, like not being successful. By addressing not only what people are ultimately striving for but also what they are ultimately afraid of, we can provide ultimate value.

CASE STUDY: CREDIT CARD THEFT

Challenge: In talking about identification and identity theft on behalf of a financial institution, we met with a group of people whose identities had been stolen. For them, it was highly emotional to remember trying to buy the house of their dreams and being rejected because someone else had fraudulently taken out another mortgage using their identity. The house was tied to much deeper underlying notions, like their "forever home" where they wanted to grow old and raise kids, as well as the negative feelings of unfair rejection they had to experience. All in all, they had a lot of fear and mistrust of the process and of financial institutions due to their past experiences.

Outcome: In each of these cases, whether it was being denied a mortgage or having a credit card rejected at Staples, these consumers had deeply emotional associations with the idea of credit. For them, it was essential that we find ways to help shape not only their unique decision-making processes, but also their perceptions and mistrust of financial institutions on the whole. In this case we found that considering the products and services in such a way that they weren't directly tied to financial institutions helped to distance them from the strong emotional experiences that could easily be elicited.

Analyzing Dreams (Goals, Life Stages, Fears)

The following case study is just one example of how dreams, goals, and fears can change by one's stage in life.

CASE STUDY: PSYCHOGRAPHIC PROFILE

Challenge: In my line of work, sometimes we create "psychographic profiles" to segment (and better market to) groups of consumers. One artificial but representative example, is shown in Figure 14-2. It relates to the questions we asked on behalf of a credit card company, which I mentioned back in Chapter 6. In these interviews, we went from the short-term emotions to the longer-term emotions to what people were ultimately trying to accomplish. These types of interviews can be like therapy (for the participants, not us).

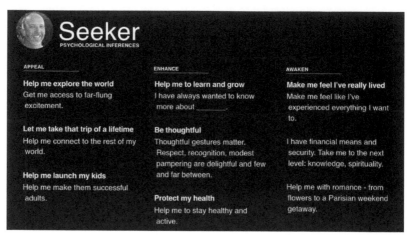

FIGURE 14-2

Example of a persona focused on what appeals to, enhances, and awakens a customer

Outcome: As you can see in the first column, the things that *appealed* to this group of consumers—older, possibly retired adults who might have grown children—were things they could do with their newly discovered free time. Things like taking that trip to Australia, or further supporting their kids in early adulthood by helping them launch their careers, buy houses, etc.

Over the course of the contextual interviews, we were able to go a bit further than the short-term goals (e.g., a trip to Australia) and get to what they were seeking to *enhance*. Things like learning to play the piano, receiving great service and respect when they stay at hotels, or maintaining/improving their health.

Then, going even longer term, we got to the notion of what they wanted to *awaken* in their lives. For many in this focus group, they were thinking beyond material success and were seeking things like knowledge, spirituality, service, or leaving a lasting impact on their community— the next level of fulfillment, awakening their deepest passions. As well as desire, we also observed a level of fear at not having these passions fulfilled. These were all emotions we wanted to address in our products and services for this group.

Getting the Zeitgeist (Person versus Persona-Specific)

In considering emotion, we're also taking into account the distinct personalities of our end audience, the deeper undercurrent of who they are, and who it is they're trying to become.

CASE STUDY: ADVENTURE RACE

Challenge: It's not every day you get to join in on a mud-filled adventure race for work. In the one pictured in Figure 14-3, many of the runners were members of the police force or former military—all very serious athletes; as you can imagine. Our client, however, saw an opportunity to attract families and "average Joes" too.

Outcome: In watching people participate in one of these races (truly engaging in contextual inquiry, covered in mud from head to toe, I might add), my team and I observed that the runners all had this amazing sense of accomplishment at the end of the race, as well as during the race. It was clear that they were each digging really deep into their psyche to push through some obstacle (be it running through freezing-cold water or crawling under barbed wire) to finish the race, and that these were metaphors for other obstacles in their lives that they might also overcome.

FIGURE 14-3
Passion at a Spartan Obstacle Race

By observing this emotional content, we saw that this was something we could harness not only for the dedicated race types, but ordinary people as well (I don't mean that in a degrading way; even this humble psychologist ran the race, so there's hope for everyone!). In our product and service design efforts going forward, we harnessed the emotional content like running for a cause (as was the case with a group of cancer survivors or ex-military who were overcoming PTSD), running with a friend for accountability, giving somebody a helping hand, or signing up with others from your family, neighborhood, or gym. We knew these deeper emotions would be crucial to people deciding to sign up and invite friends.

A Crime of Passion (In the Moment)

Remember the concept of *satisficing*? It's somewhere between *satisfactory* and *sufficing*. Satisficing is all about emotion, and defaulting to the easiest or most obvious solution when we're overwhelmed. There are many ways we do this, and our interactions with digital interfaces are no exception.

Maybe a web page is too busy, so we satisfice by leaving the page and going to a tried-and-true favorite. Maybe we're presented with so many options for a product (or candidates on the ballot for our state's primary elections?) that we just select the one that's displayed the most prominently, not taking specifications into consideration. Maybe we buy something out of our price range simply because we're feeling stressed out and don't have time to keep searching. You get the idea.

CASE STUDY: MAD MEN (AND WOMEN)

Challenge: Let me tell you a story about a group of young ad executives with whom we did contextual inquiry. Their first job out of college was with a prestigious ad agency in downtown New York City. They thought it was all very cool and were excited about their career possibilities. Often, they were put in charge of buying a lot of ads for a major client, and they literally were tasked with spending $10 million on ads in one day. In observing this group of young ad-men and – women, we saw emotions were running high. They were fearful because this task— picking which ads to run and on which stations—if done incorrectly, could end not only their careers, but also the "big city ad executive" lifestyles and personas they had cultivated for themselves. Making one wrong click would end their whole dream and send them packing in shame (or so they believed). There was an analytics tool in place for this group of ad buyers, but because they were so nervous, they would default to old habits (satisfice) and simply wouldn't trust the automated system (even if it outperformed their own performance).

Outcome: We tweaked the analytics tool to show all the ad campaign stats the buyers would need at a glance. We made them very simple to understand and used visual attributes like bar graphs and colors to grab their attention. With so many emotions weighing in on their decisions, we wanted to make sure this tool made it clear what needed to be done next.

Real-World Examples

Let's take a look at the sticky notes relevant to emotion (Figures 14-4 through 14-7):

"Loves that the product reviews are sorted by popularity."

Comments that use words like "love" or "hate" shouldn't necessarily be grouped into emotion, since their meaning always depends on context. This comment is talking about a specific design feature, so you could consider vision, wayfinding, or even memory if it's meeting an expectation the user had. I'm still a bit torn, but I think the emotion of delight could be the strongest factor in this case.

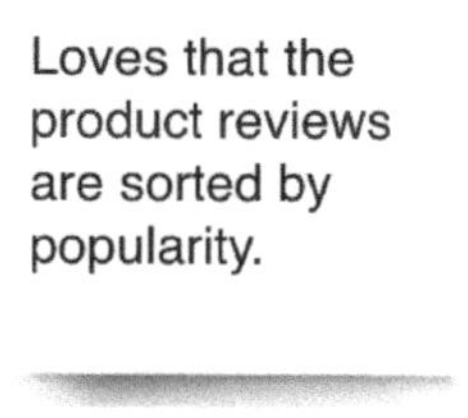

FIGURE 14-4

Loves that the product reviews are sorted by popularity.

Research observation: emotional terms do not always suggest an emotional source to response

"Wants his clothes to hint at his position (Senior Vice President)."

The comment we just looked at related to an immediate emotion tied to a specific stimulus. This one, however, exemplifies the deeper type of emotion I've been talking about (albeit perhaps with more possibility for nobility). Sure, you could argue it's more of a surface-level comment—that this person is merely browsing for a certain type of clothing. But I would argue it speaks to a deep-seated desire to display a certain persona or image, and be perceived by others in that light. I think it encompasses a desire to look powerful, be treated a certain way, drive a certain type of car—or whatever this person envisions as representing "success."

FIGURE 14-5

Wants his
clothes to hint
at his position
(Senior Vice
President)

Research observation:
hints at deeper
motivations for
decision

"Says 'reviews are a scam, the store makes them up. I don't trust them!'"

This one reads like pure emotion to me. Credibility and trust when it comes to ecommerce sites seem to be big hurdles for this user. What are some ways to present reviews such that these fears would be allayed?

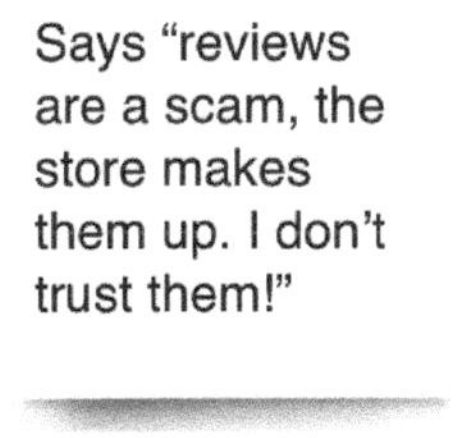

FIGURE 14-6

Says "reviews
are a scam, the
store makes
them up. I don't
trust them!"

Research observation:
strong emotional
response could be
caused by something
outside the product

[SIDE NOTE]

Remember to take your users' feedback in totality. In addition to this comment about reviews, this customer also remarked that he was afraid of "getting burned again" and that he wanted a way to compare products the way *Consumer Reports* does. Taken in totality, we can surmise that this person might have issues working up trust for any ecommerce system. With feedback like this, we want to think about what we could do to make our site trustworthy for our more nervous customers.

"Afraid she might buy something by accident."

This comment doesn't include anything specific about interaction issues leading to this fear of accidentally buying something, so I would label it as an immediate emotional consideration, rather than anything more deep-seated.

FIGURE 14-7
Research observation: straightforward emotional comment

Afraid she might buy something by accident.

Concrete Recommendations

- Build up a set of questions throughout the interview that systematically go from the innocuous to the most meaningful (e.g., What credit cards are in your wallet? What do you like to do on the weekends? What makes you happiest? What are your goals this year? What would make you a success? What are you most fearful about that would stop you from getting there?).

- Determine how the customer's goals in this case (e.g., shopping for clothes) fit into the larger picture of their life (e.g., excited to find someone to marry, want to feel young again, want to feel professional and be taken seriously).

- When building a persona, ensure that life stage and major fears are captured (e.g., older and fearful of having to search for a new job). Fears are powerful drivers away from logic.

- Estimate how much is on the line with the decision the customers are making (picking the right gum won't get you fired, but getting the analysis wrong for the election will).

Putting the Six Minds to Work in Your Designs

When I was studying for my PhD, I had a very distinguished professor who, two and a half hours into a three-hour seminar would say, "So what?! So...what?!" This evoked terror in all the graduate students, but really his goal was to say—in a somewhat confrontational way—"Why should I care? How can I use this?"

Part III is all about the "So what?" of the Six Minds. What do we do with the data once we've collected it? What do you get out of it? How can you go from a piece of evidence to insight? And how should these insights influence your product and service designs? How can we use this knowledge to benefit and influence your customers?

Put simply, how can all this knowledge about, and understanding of, end users enable product and service designers to design better products and services?

In the following chapters, we'll look at how we can apply our data on the Six Minds, determine if our current interface is working, and design differently to match our audience's ideal experience. We'll also look at real-life examples of where cognitive design has influenced digital design.

Last, we'll look at how you fit all of this into your everyday process. I've been in the trenches with developers, so trust me. I've got ammunition for those of you who think your bosses or developers won't go for this approach.

[15]

Sense-Making

IN PART II, WE collected data from contextual inquiry interviews and sorted it into our Six Minds framework. Now we can move onto our next major goals:

- Looking for commonalities among dimensions of the Six Minds (level of expertise, feelings of anxiety, etc.)

- Segmenting customers by their needs (e.g., novices versus experienced professionals, supervisors versus analysts, parents versus children) and relevant dimensions (e.g., word usage, microgoals, underlying assumptions), and building a psychographic profile of each segment

Finally, we'll end this chapter with a few notes about another system of classification (See/Feel/Say/Do) and why I believe it fails to organize the data in a way that is truly helpful for product and service design.

Affinities and Psychographic Profiles

Using our Six Minds framework, let's review the findings for our example participants from Part II. We have all our sticky notes grouped by participant and by the Six Minds. Now it's time to look at these findings and see if there are any relationships between them or any underlying similarities in how these individuals are thinking (Figure 15-1).

1

Decision Making	Language	Emotion	Memory	Wayfinding	Vision
Concern: "What if the $5,000 chair I'm buying is damaged in transit?"	Searched for "Eames Midcentury Lounge Chair" when asked to search for a chair.	Wants his clothes to hint at his position (Senior Vice President)	Expected the search to provide type-ahead choices.	Expected that clicking on a book cover would reveal its table of contents.	Can't find the "Save for Later" feature.
Loves that the product reviews are sorted by popularity.	Couldn't find the shopping cart. Eventually figured out the "Shopping bag" is the "cart"			Can't seem to find the button to play a movie preview.	

2

Decision Making	Language	Emotion	Memory	Wayfinding	Vision
Wants to know right away if this site will accept Paypal.	Searched for "Dewalt 2-speed 20 volt cordless drill"	Worried about "getting burned again". Wants return policy before clicking on "Add to bag"	Wants a way to compare products side by side like Consumer Reports	Can't figure out how to get zoomed in the pictures of the product.	Didn't notice "Return to results" link. Looking for a "back" button
		Says "reviews are a scam, the store makes them up. I don't trust them!"		Clicked store logo. Couldn't figure out how to get back to search results	

3

Decision Making	Language	Emotion	Memory	Wayfinding	Vision
Wants easy way to buy on laptop and send movie to TV	Wants to know if movie plays in "1080p or 4K UHD"		Expects to see "Rotten Tomato" ratings for movies.	Expects that clicking on the movie starts a preview, not actual movie.	Movie listings seem really busy to him with a lot of words.
				Wants to filter results by "Film noir"	Can't see which movies are included in a membership.

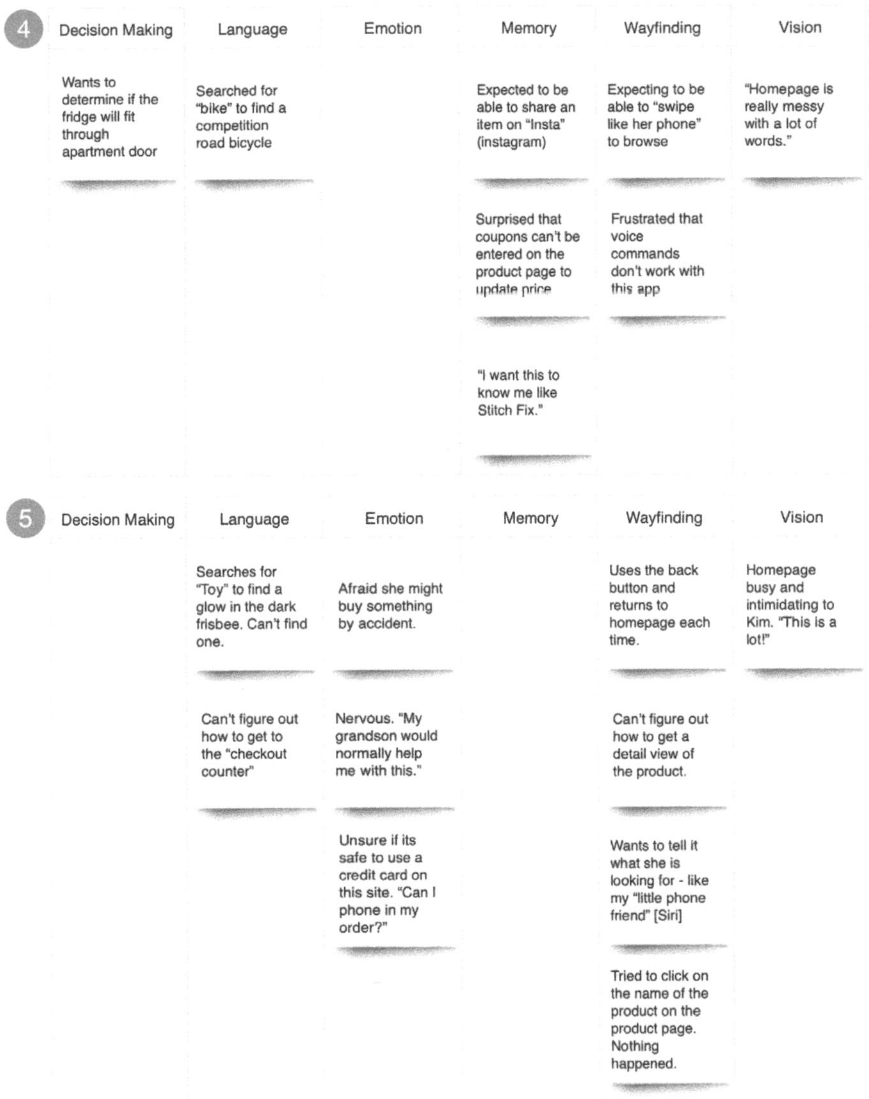

4 Decision Making	Language	Emotion	Memory	Wayfinding	Vision
Wants to determine if the fridge will fit through apartment door	Searched for "bike" to find a competition road bicycle		Expected to be able to share an item on "Insta" (instagram)	Expecting to be able to "swipe like her phone" to browse	"Homepage is really messy with a lot of words."
			Surprised that coupons can't be entered on the product page to update price	Frustrated that voice commands don't work with this app	
			"I want this to know me like Stitch Fix."		

5 Decision Making	Language	Emotion	Memory	Wayfinding	Vision
Searches for "Toy" to find a glow in the dark frisbee. Can't find one.		Afraid she might buy something by accident.		Uses the back button and returns to homepage each time.	Homepage busy and intimidating to Kim. "This is a lot!"
Can't figure out how to get to the "checkout counter"		Nervous. "My grandson would normally help me with this."		Can't figure out how to get a detail view of the product.	
		Unsure if its safe to use a credit card on this site. "Can I phone in my order?"		Wants to tell it what she is looking for - like my "little phone friend" [Siri]	
				Tried to click on the name of the product on the product page. Nothing happened.	

FIGURE 15-1

Findings from contextual interviews, organized by the Six Minds

Language

Looking at the column for Language, I see that Participants 1, 2, and 3 are all using terms like "Eames Midcentury Lounge Chair," or "DeWalt 2-speed, 20-volt cordless drill," or "1080p or 4K UHD" (Figure 15-2).

While they're searching for very different things, these three participants all have pretty sophisticated language for what they're talking about. They seem to be experts, if not professionals, in their particular fields, and are very knowledgeable in the subject matter.

**① **

Decision Making	Language	Emotion	Memory	Wayfinding	Vision
Concern: "What if the $5,000 chair I'm buying is damaged in transit?"	Searched for "Eames Midcentury Lounge Chair" when asked to search for a chair.	Wants his clothes to hint at his position (Senior Vice President)	Expected the search to provide type-ahead choices.	Expected that clicking on a book cover would reveal its table of contents.	Can't find the "Save for Later" feature.
Loves that the product reviews are sorted by popularity.	Couldn't find the shopping cart. Eventually figured out the "Shopping bag" is the "cart"			Can't seem to find the button to play a movie preview.	

Experts ←

**② **

Decision Making	Language	Emotion	Memory	Wayfinding	Vision
Wants to know right away if this site will accept Paypal.	Searched for "Dewalt 2-speed 20 volt cordless drill"	Worried about "getting burned again". Wants return policy before clicking on "Add to bag"	Wants a way to compare products side by side like Consumer Reports	Can't figure out how to get zoomed in the pictures of the product.	Didn't notice "Return to results" link. Looking for a "back" button
		Says "reviews are a scam, the store makes them up. I don't trust them!"		Clicked store logo. Couldn't figure out how to get back to search results	

**③ **

Decision Making	Language	Emotion	Memory	Wayfinding	Vision
Wants easy way to buy on laptop and send movie to TV	Wants to know if movie plays in "1080p or 4K UHD"		Expects to see "Rotten Tomato" ratings for movies.	Expects that clicking on the movie starts a preview, not actual movie.	Movie listings seem really busy to him with a lot of words.
				Wants to filter results by "Film noir"	Can't see which movies are included in a membership.

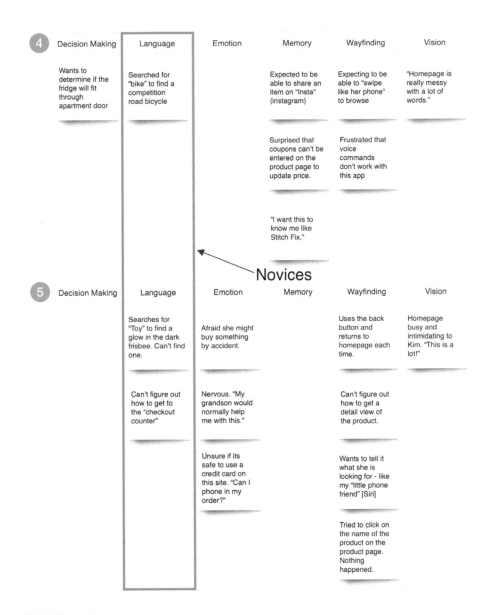

4 Decision Making	Language	Emotion	Memory	Wayfinding	Vision
Wants to determine if the fridge will fit through apartment door	Searched for "bike" to find a competition road bicycle		Expected to be able to share an item on "Insta" (instagram)	Expecting to be able to "swipe like her phone" to browse	"Homepage is really messy with a lot of words."
			Surprised that coupons can't be entered on the product page to update price.	Frustrated that voice commands don't work with this app	
			"I want this to know me like Stitch Fix."		

Novices

5 Decision Making	Language	Emotion	Memory	Wayfinding	Vision
	Searches for "Toy" to find a glow in the dark frisbee. Can't find one.	Afraid she might buy something by accident.		Uses the back button and returns to homepage each time.	Homepage busy and intimidating to Kim. "This is a lot!"
	Can't figure out how to get to the "checkout counter"	Nervous. "My grandson would normally help me with this."		Can't figure out how to get a detail view of the product.	
		Unsure if its safe to use a credit card on this site. "Can I phone in my order?"		Wants to tell it what she is looking for - like my "little phone friend" [Siri]	
				Tried to click on the name of the product on the product page. Nothing happened.	

FIGURE 15-2

Reviewing the commonalities between participants within the Language section of the Six Minds

In contrast, it looks like Participant 4 searched for a "bike" while looking for a competition road bike. Participant 5 searched for a glow-in-the-dark Frisbee by typing in "toy." Participant 5 also talked about getting to the "checkout counter," rather than "Amazon checkout" or "QuickPay," or something else that would suggest more knowledge of how the online shopping experience works.

Just by looking at the language, we can see that we've got some people with expertise in the field, and others who are much more novices in the area of ecommerce. Moving forward, it might make sense to examine how the experts approach aspects of the interface, versus how the novices approach those same aspects, and see if there are similarities across those individuals.

This is just a start, though. We don't want to pigeonhole anyone into just one category, because ultimately we are trying to find commonalities on many dimensions. Using a different dimension or "mind," we could look at these same participants in a very different way.

Let's take emotion (Figure 15-3). Looking across all the findings, we see that Participants 2 and 5 both seemed pretty concerned about the situation, and were afraid that something bad might happen or that they would "get burned again." We're definitely seeing some uncertainty and reticence to go ahead and take the next step because they're worried about what might happen. These folks might need some reassurance. Participants 1, 3, and 4, on the other hand, aren't displaying any of that emotion or hesitation.

Could there be other areas in which Participants 2 and 5 are also on the same page? Maybe there are similarities in how they do wayfinding, or the information they're looking for, as opposed to Participants 1, 3, and 4, who seem to be going through this process in a more matter-of-fact way.

Emotion

1

Decision Making	Language	Emotion	Memory	Wayfinding	Vision
Concern: "What if the $5,000 chair I'm buying is damaged in transit?"	Searched for "Eames Midcentury Lounge Chair" when asked to search for a chair.	Wants his clothes to hint at his position (Senior Vice President)	Expected the search to provide type-ahead choices.	Expected that clicking on a book cover would reveal its table of contents.	Can't find the "Save for Later" feature.
Loves that the product reviews are sorted by popularity.	Couldn't find the shopping cart. Eventually figured out the "Shopping bag" is the "cart"			Can't seem to find the button to play a movie preview.	

4

Decision Making	Language	Emotion	Memory	Wayfinding	Vision
Wants to determine if the fridge will fit through apartment door	Searched for "bike" to find a competition road bicycle		Expected to be able to share an item on "Insta" (instagram)	Expecting to be able to "swipe like her phone" to browse	"Homepage is really messy with a lot of words."
			Surprised that coupons can't be entered on the product page to update price.	Frustrated that voice commands don't work with this app	
		"I want this to know me like Stitch Fix."			

← ———— Neutral emotions

3

Decision Making	Language	Emotion	Memory	Wayfinding	Vision
Wants easy way to buy on laptop and send movie to TV	Wants to know if movie plays in "1080p or 4K UHD"		Expects to see "Rotten Tomato" ratings for movies.	Expects that clicking on the movie starts a preview, not actual movie.	Movie listings seem really busy to him with a lot of words.
				Wants to filter results by "Film noir"	Can't see which movies are included in a membership.

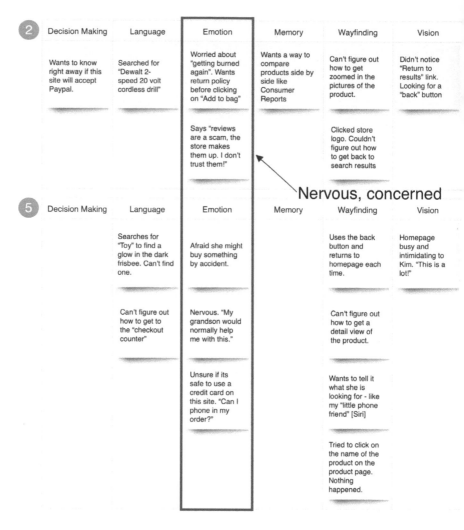

② Decision Making	Language	Emotion	Memory	Wayfinding	Vision
Wants to know right away if this site will accept Paypal.	Searched for "Dewalt 2-speed 20 volt cordless drill"	Worried about "getting burned again". Wants return policy before clicking on "Add to bag"	Wants a way to compare products side by side like Consumer Reports	Can't figure out how to get zoomed in the pictures of the product.	Didn't notice "Return to results" link. Looking for a "back" button
		Says "reviews are a scam, the store makes them up. I don't trust them!"		Clicked store logo. Couldn't figure out how to get back to search results	

Nervous, concerned

⑤ Decision Making	Language	Emotion	Memory	Wayfinding	Vision
	Searches for "Toy" to find a glow in the dark frisbee. Can't find one.	Afraid she might buy something by accident.		Uses the back button and returns to homepage each time.	Homepage busy and intimidating to Kim. "This is a lot!"
	Can't figure out how to get to the "checkout counter"	Nervous. "My grandson would normally help me with this."		Can't figure out how to get a detail view of the product.	
		Unsure if its safe to use a credit card on this site. "Can I phone in my order?"		Wants to tell it what she is looking for - like my "little phone friend" [Siri]	
				Tried to click on the name of the product on the product page. Nothing happened.	

FIGURE 15-3

Reviewing the commonalities between participants within the Emotion section of the Six Minds

Using different dimensions, we can look at people and see how we might group them. Ideally, we would hope to find similarities across multiple dimensions. I'm using a tiny sample size for the sake of illustration in this book, but typically, we would be looking at this with a much broader set of data—perhaps 24 to 40 people—anticipating groups of 4 to 10 people, depending on how the segments fall out.

Wayfinding

When we look at wayfinding we see that Participants 1, 2, 3, and 5 all had problems with the user experience or the way that they interacted with a laptop (Figure 15-4). Participant 4 approached the experience very differently, expressing a desire to be able to "swipe like her phone" or use voice commands. This participant seems far more familiar with the technology, to the extent that she has surpassed it and is ready to take it to the next level. Through the Wayfinding lens, we see that our participants are looking at the same interface using varied tools and with varied expectations in terms of their level of interaction design and sophistication.

1

Decision Making	Language	Emotion	Memory	Wayfinding	Vision
Concern: "What if the $5,000 chair I'm buying is damaged in transit?"	Searched for "Eames Midcentury Lounge Chair" when asked to search for a chair.	Wants his clothes to hint at his position (Senior Vice President)	Expected the search to provide type-ahead choices.	Expected that clicking on a book cover would reveal its table of contents.	Can't find the "Save for Later" feature.
Loves that the product reviews are sorted by popularity.	Couldn't find the shopping cart. Eventually figured out the "Shopping bag" is the "cart"			Can't seem to find the button to play a movie preview.	

2

Decision Making	Language	Emotion	Memory	Wayfinding	Vision
Wants to know right away if this site will accept Paypal.	Searched for "Dewalt 2-speed 20 volt cordless drill"	Worried about "getting burned again". Wants return policy before clicking on "Add to bag"	Wants a way to compare products side by side like Consumer Reports	Can't figure out how to get zoomed in the pictures of the product.	Didn't notice "Return to results" link. Looking for a "back" button
		Says "reviews are a scam, the store makes them up. I don't trust them!"		Clicked store logo. Couldn't figure out how to get back to search results	

5

Decision Making	Language	Emotion	Memory	Wayfinding	Vision
	Searches for "Toy" to find a glow in the dark frisbee. Can't find one.	Afraid she might buy something by accident.		Uses the back button and returns to homepage each time.	Homepage busy and intimidating to Kim. "This is a lot!"
	Can't figure out how to get to the "checkout counter"	Nervous. "My grandson would normally help me with this."		Can't figure out how to get a detail view of the product.	
		Unsure if its safe to use a credit card on this site. "Can I phone in my order?"		Wants to tell it what she is looking for - like my "little phone friend" [Siri]	
				Tried to click on the name of the product on the product page. Nothing happened.	

Trouble with computer UX →

3

Decision Making	Language	Emotion	Memory	Wayfinding	Vision
Wants easy way to buy on laptop and send movie to TV	Wants to know if movie plays in "1080p or 4K UHD"		Expects to see "Rotten Tomato" ratings for movies.	Expects that clicking on the movie starts a preview, not actual movie.	Movie listings seem really busy to him with a lot of words.
				Wants to filter results by "Film noir"	Can't see which movies are included in a membership.

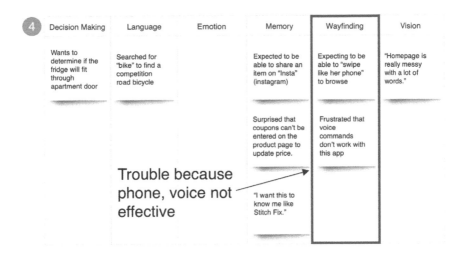

4	Decision Making	Language	Emotion	Memory	Wayfinding	Vision
	Wants to determine if the fridge will fit through apartment door	Searched for "bike" to find a competition road bicycle		Expected to be able to share an item on "Insta" (instagram)	Expecting to be able to "swipe like her phone" to browse	"Homepage is really messy with a lot of words."
				Surprised that coupons can't be entered on the product page to update price.	Frustrated that voice commands don't work with this app	
		Trouble because phone, voice not effective		"I want this to know me like Stitch Fix."		

FIGURE 15-4

Reviewing the commonalities between participants within the Wayfinding section of the Six Minds

We've seen three ways we could group this batch of users, and there are others. Which one does it make the most sense to pick? Sometimes—actually, it's pretty common—it might be really obvious to you that certain people are of one accord and can be grouped together. Other times, you may see some commonalities among a subgroup, but there's no real equivalent from the others.

Finding the Dimensions

The best way to illustrate the audience segmentation process is through real-life examples. With the case studies that follow, I'll show you just one grouping per data set, to give you a taste of the kinds of groupings you might produce.

CASE STUDY: MILLENNIAL MONEY

When working with a worldwide online payments system (you've heard of it), my team and I performed contextual interviews with Millennials, trying to understand how they use and manage money and how they define financial success.

The sticky notes in Figure 15-5 represent real-life findings from people we met with. Once again, the question is: "How would I organize these folks?" This picture shows all the notes sorted by person for a handful of the participants.

FIGURE 15-5
Participants displayed with relevant data in each vertical

One subset of the participants shared the commonality of a life focused on adventure. Their desired lifestyle dramatically affected their decision making about money—after saving just enough, they would immediately use that money for their latest Instagram-ready adventure. We saw that what made these participants happy, and their deepest goal (emotion), was to have new experiences and adventures. They were putting all their time and money into adventures and travel. Ultimately, what was really meaningful to them at that deep emotional level was experiences, rather than things.

Another observation that falls within the category of emotion is that the participants weren't really defining themselves by their job or other traditional definitions (e.g., "I'm an introvert"); instead, they seemed to find their identity in the experiences they wanted to have.

Socially speaking, these people were instigators, and tried to get others to join them in their adventures. New possibilities, and offers of cheap tickets for which they knew the ticket codes (language), easily attracted their attention (attention!). They were masters of manipulating social networks (Instagram, Pinterest) to share and learn about new places (wayfinding, perhaps both in our sense of wayfinding—the ability to navigate apps like Instagram—and physical wayfinding to find these new places!).

Using the Six Minds framework, I think we saw several things in these participants (remember, this was on behalf of that big online payments system, so we were particularly interested in groupings related to how participants managed their money):

Decision Making

They weighed every money-related decision against their goal of having new experiences. Their level of commitment to long-term financial planning was nonexistent because they were focused on living in the moment.

Emotion

Maximizing their adventure-readiness was paramount, and anything that got in the way of that happiness was seen as negative. Conversely, any payment system that helped them achieve their far-flung lifestyle was a plus.

Language

They had amazing vocabulary and knowledge of travel sites and airline specials—even the codes for airfares (did you know airfares have codes?). They spoke the lingo of frequent flyer miles, baggage fees, and rental cars because they were experts in travel.

We looked at other audiences for that study, but I just wanted to give you a feeling for a real grouping and how in this case we mostly used decision making, emotion, and a bit of language as the key drivers for how we organized participants. Other dimensions, like the way that they interacted with apps (wayfinding) or their underlying experiences and the metaphors they used (memory), just weren't as important as the central concept of adventure that they organized their lives around.

This is pretty common. With audience segmentation, higher-order dimensions, like decision making and emotion, tend to stand out more than the other Six Minds dimensions. If your study is focused on an interface-intensive design situation, like designers working on the program Adobe Photoshop, for example, you might find groupings that have more to do with vision or wayfinding.

CASE STUDY: TRUST IN CREDIT

The second example I want to share concerns a study we did on behalf of a top 10 financial company. We studied how much people know about their credit scores and how they're affected by them. We also wanted to gauge people's overall knowledge about credit and fraud.

We counned several types of groups, but I'm only going to describe one persona here: the Fearful & Unsure group. For them, financial transactions were tied much more to emotion than is typical of the average individual. One woman who we'll call Ruth was incredibly embarrassed when her credit card was denied at the grocery store because someone had actually stolen her identity. This led to feelings of anxiety, denial, powerlessness, fear, and being overwhelmed.

In this audience segment, people like Ruth just hid from the issue (Emotion). Unlike other people who were inspired to take action and learn about credit to protect themselves, members of this Fearful & Unsure segment were too shell-shocked to take much action (Decision Making). They tried to avoid situations the problem could occur again, and were operating in a timid defense mode, rather than offense (Attention). This Fearful & Unsure group didn't consider themselves credit-savvy, which was consistent with the way they spoke about credit issues (Language).

What's driving this psychographic profile is the individuals' emotional response to a situation involving credit, resulting in a unique pattern of decision making, the language of a novice, and an unawareness of ways to reduce their credit risk in the future.

Challenging Internal Assumptions

In the examples we just reviewed, I used more complex cognitive processes like decision making, emotion, and language to segment audiences across an industry or target audience pool. In many cases, you might have a boss or a manager who's used to doing audience segmentation in a different way (e.g., "We need people who are in these age ranges, spend this much, are in these socio-economic brackets, or have these titles"). I want you to be ready to challenge some of these assumptions.

If your analysis seems to contradict some of the big patterns from yesteryear, be prepared to receive some pushback. Don't be afraid to say, "No, our data is actually inconsistent with that"—and point to the data! Go ahead and test the old assumptions to see if they're still valid.

When possible, try to get a sample that's at least 24 people. If you can get a geographically dispersed or even linguistically diverse sample, so much the better. All of these considerations are ammunition to help you answer the question, "Was this an unusual sample?" When you have a large, diverse sample, you can say, "No; this is well beyond one or two people who happen to think in this way."

At times, you'll need to challenge not only the outdated notions of colleagues, but also your own preconceptions. Sometimes the data we find challenges our own ideas and ways of organizing material. In these cases, I urge you to make sure that you're representing the data accurately, and not viewing it through the lens of presupposition.

Back in Chapter 7, I challenged you to approach contextual inquiry with a "tabula rasa" (clean slate) mentality. Leave your assumptions at the door and be open to whatever the data says. The same goes for audience segmentation. As best you can, try not to approach the data with your own hypotheses. We know statistically that people who say "I know that X is true" are looking for confirmation of X in the data, as opposed to people who are just testing out different possibilities. Be the latter type of analyst. Be open to different possibilities.

Always try to negate the other possibilities, as opposed to only looking for information that confirms your hypothesis. Look carefully at the underlying emotional drivers for each person. How are they moving around their problem space, and what are they finding out? What are the past experiences that are affecting them, and do the segments hinge on those past experiences?

In the case of the small business owners mentioned in Chapter 12, we went back and forth a lot on what the salient features were for audience segmentation. There was the fact that they problem-solved from two very different perspectives. There was the language and sophistication level factor, which also differed greatly based on the subject matter (e.g., craft expertise versus business expertise). There were several patterns we were seeing. To land on our eventual segmentations, we had to test each of these patterns across our audience segmentations and make sure those patterns were really borne out across the data.

Finally, when possible, try to organize your participants by the highest-level dimension possible. Even though you may start out with surface-level observations, try to go deeper to get at those drivers and the underlying goals. As a psychologist, I give you permission to dig deeper into the psyche—into the big, bold emotional state that influences the way people make decisions.

Ending an Outdated Practice: See/Feel/Say/Do

If you are familiar with empathy research, you may have heard of the See/Feel/Say/Do chart (Figure 15-6). These charts are popular with many groups looking for a mechanism to develop empathy for a user. They may also be used to help to segment customers into groupings.

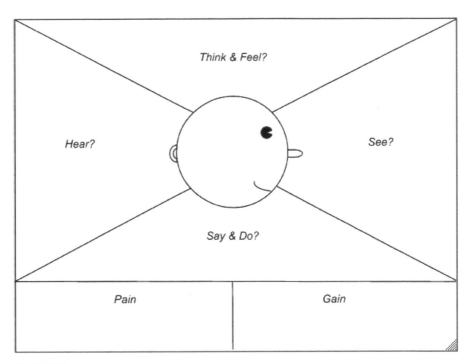

FIGURE 15-6
Customer empathy map

A lot of empathy research programs use a diagram like this one. See/Feel/Say/Do diagrams ask the following questions of your customers:

- "What are they seeing?"
- "What are they feeling and thinking?"
- "What are they saying and doing?"
- "What are they hearing?" (some diagrams leave this one out)

The diagram shown here also includes a Pain/Gain component, asking "What are some of the things your user is having trouble with?" (pain) and "What are some opportunities to improve those components?" (gain).

To some folks, this looks a lot like the Six Minds. What's the difference? Let's look more closely at how they align:

See

> At first glance, this is pretty clearly tied to vision. But remember: when we consider what the user is seeing, we want to know what they're actually looking at or attending to—not necessarily what we're presenting to them, which may be different. I want to make sure that we're thinking from an audience perspective, and taking into account actually what they're really seeing. There is an important component that is missing here: What are they *not* seeing? We want to know what they are searching for, and why. It is crucial to consider the attentional component, too.

Feel

> It may seem like this would be synonymous with emotion— think again. In these empathy diagrams, "Feel" is talking about the immediate emotion of a user in relation to a particular interface. The questions is, "What is the user experiencing at this very moment?" As you may remember from our discussion of emotion, from a design perspective, we should be more interested in the deeper, underlying sources of emotion. What is it that your users are trying to achieve? Why? Are they fearful of what might happen if they're not able to achieve that? What are some of their deepest concerns? In other words, we want to push beyond the surface level of immediate reactions to an interface and consider the more fundamental concerns that may be driving those reactions.

Say

> Often, saying and doing are grouped together, but first let's focus on the saying. I struggle with the notion of just reporting what users are actually saying, because at the end of the day, they can say things that relate to any of the Six Minds. They might say what they're trying to accomplish, which would be decision making. They could describe how they're interacting with something, which would be wayfinding. Or maybe they're saying something about emotion. If you recall the types of observations we grouped under Language in Part II with our sticky note exercise, they weren't a mere litany of all the things our users uttered; we identified all the things our users said (or did) that pertained to the words being

used in a certain interface. I believe that language has to do with the user's level of sophistication and the terms they expect to see when interacting with a product.

By categorizing all of the words coming out of our users' mouths as "Say," I worry that we're oversimplifying things and missing what those statements are getting at. I also think we miss the chance to determine people's overall expertise level based on the kinds of words they're using. This grouping doesn't lend itself to an organization that can influence product or service design.

Do

When we think about wayfinding, that has to do with how people are getting around in the interface or using the service. We're asking questions like, "Where are you in the process?" and "How can you get to the next step?" I think wayfinding gets at what the user is actually doing, as well as what they believe they can do, based on their perception of how this works. Just describing behavior ("Do") is helpful, but not sufficient.

Decision Making (missing)

While "Do" comes close, I think that we're missing decision making in See/Feel/Say/Do. This style of representation doesn't mention how users are trying to solve their problem. With decision making or problem solving, we want to consider how our users think they can solve the problem and what operators they think they can use to move around their decision space.

Memory (missing)

Memory also gets short shrift in the See/Feel/Say/Do model. We want to know the metaphors people are using to solve their problem, and the expected interaction styles they're bringing into this new experience. We want to know about their past experiences and expectations, including those that users might not even be aware of, but that they imply through their actions and words (like how they expect the experience of buying a book to work based on how Amazon works, or how they expect a sit-down restaurant to have waiters and white tablecloths). See/Feel/Say/Do charts leave out the memories and frameworks users are employing.

Hopefully you can see that while it's better than nothing, See/Feel/Say/Do is missing key elements that we need to consider, and that it also oversimplifies some of the pieces it does consider. We can do better with our Six Minds.

Concrete Recommendations

- Segment audiences using the following techniques:

    - Collect findings from the contextual interviews across the Six Minds

    - Determine which participants and activities share "affinities" or commonalities

    - Determine how users differ in their needs, segmenting them by relevant dimensions

[16]

Putting the Six Minds to Work: Appeal, Enhance, Awaken

At this point we have completed contextual interviews, extracted interesting data points from each participant, and organized them according to the Six Minds. We've also segmented the audience into different groupings. It's time to put that data to work and think about how it should influence our products and services.

In this chapter I'll explain exactly what I mean by appealing to your audience, enhancing their experience, and awakening their passion. A summary might be:

- What would appeal to a user immediately so they buy into the product or service?

- What aspects of the product or service would keep a customer happy by providing utility and a great long-term experience?

- How might the product or service help users realize their deepest goals and desires?

We'll look at the type of Six Minds data that we would use to do those things. And as usual, I'll give you some examples of how to put all of this into practice.

Appeal: What People Say They Want

Our first focus is all about appealing to your customers. A digital reference point for this section is the popular website Cool Hunting (*https://coolhunting.com*). If you're not familiar with the site, it essentially curates articles and recommendations on everything from trendy hotels to yoga gear to the latest tech toys. What are the hot new trends that people are looking for and saying that they want?

You might think this is a bit superficial given our discussions of going beyond trends and getting at your customers' underlying desires. But regardless of what you're offering, it's absolutely essential that you can attract them. They need to feel like your product or service is what they are looking for.

Some of the time, what our customers want and what might benefit them are two very different things. The task before us in that case isn't easy. We have to be prepared to attract them and appeal to them using what they *think* they want—even if we know that something else might be much more beneficial for them. Ideally, after attracting them we can educate them better in order to help the customers make an informed decision.

We have to start with where people are, and what they're *saying* they want and need, even if that's not necessarily the case deep down. We start by taking what they're saying at face value.

Going to our Six Minds data, I think there are three dimensions that are most relevant to this conversation around appeal:

Vision/Attention

As you watch people interacting with digital products and services, what is it that they're looking for visually? They might be looking for certain images, words, or charts. Maybe they're scanning a specific section of a website, tool, or app and expecting a particular feature. You want to ask yourself what are they looking for, and why.

Language

Your users might be looking for specific words. In this case, take into account the actual words they're using to describe what they're seeking. A customer may be looking for a "balance transfer," even though what might be much more beneficial to them as an individual would be "credit counseling."

Decision Making

Also consider the problem your customers believe they are trying to solve, and what they view as the solution. As I've mentioned, it's certainly possible that their ultimate problem has a different root cause than what they're thinking of. Someone with a cough, for example, might assume cough medicine is the solution, when the ultimate problem actually might be related to allergies. Before we can offer users with what they really need, we have to first identify their perceived problem and solution.

Enhance: What People Need

Next, we'll look at how you go one step further to enhance your customers' lives. In keeping with the popular website theme, consider Lifehacker (*https://lifehacker.com/*), a site that directs you to DIY tips and general life advice for the modern do-er. It's a great example of presenting an audience with things that actually meet their needs and solve their problems. To enhance our users' lives, we need to go a bit beyond what they're saying to consider what would really solve a problem of theirs:

Longer-term solutions

Think of Uber or Lyft. People were having a hard time getting taxis late in the evening. Maybe the taxi didn't show up, or they experienced poor customer service, or what they really needed was an easy way to order a taxi in advance or arrange one for their mom. Through a new type of tool, you might be able to present users with a longer-term solution for their problem. Now there are new companies offering even more specialized rides, like car services for young children or elderly adults, where you can track where they are, say hello, and see how they are doing along the way!

Totally new services

Your audience might need a novel type of reminder or alert system. Maybe they jot things down in a notepad, but they don't refer to that throughout the day, so they end up forgetting to pick up groceries on the way home. They're on their phones all the time, so what would be more effective is having those reminders or notes on a phone.

Teaching a function

Perhaps your audience are wasting time seraching for particular emails that they know are in their inboxes, but just can't find. A solution might be teaching them a shortcut or command like typing in a colon and the sender's email address to only pull up emails from that person. There might be ways to teach users a function that would work better for them or save them hours of work.

Novel tools

Maybe your customers are trying to meet someone in person but their schedules just aren't aligning. It might be better if you taught them how to use a video chat tool.

These are all examples of things that could change people's behavior, save them a significant amount of time, and address a concrete problem in the reasonably near future. Now let's look at how our Six Minds apply:

Decision Making

You probably won't be surprised to learn that we'll focus on the notion of decision making and look at the challenges our customers are facing. What are some other pain points they've got right now, maybe in their business, or their commute? Why are they having this problem? Is it that the transit system is bad or is it because they haven't learned to use the transit system? To come up with a solution for our users, we have to identify their true problem. This is very similar to the concept we discussed earlier of design thinking. If you remember, with design thinking, we're talking about empathy research and how it can help us understand what the problem really is.

Memory

We also want to know if our audience's frame of reference is up to date with modern technologies and tools. It might be that our users are trying to figure out an easier way to mail a check, when really it might be better for them to learn how to pay a bill online. Often I'm working on digital solutions, so I think about how we can do things in an online world that might be different from people's paper-and-pencil worldview, or even a more traditional online worldview (i.e., email versus text message or video conferencing or AI). One example of this is new tools that allow you to ask someone else to have a meeting with you. The tool automatically responds and says, "John is busy on Tuesday, but how about Wednesday?" There are many times when people's frame of reference might be based on an outdated way of doing things, rather than an understanding of the modern tools available to them.

Emotion

Many times, the challenges people encounter are tied to strong emotions. We want to look at what the pain drivers are in these situations. What might be causing your audience to be upset or dissatisfied with the current solution? What's the underlying source of that feeling? Going back to our taxi/Uber example, the problem might actually lie in the fact that they're worried about the timing, reliability, or safety of their ride. If we know the issue has to do with personal safety, then we can find a solution that addresses and overcomes that fear. We also want to know what specifically is causing the pain point—e.g., the user needs assurance that they'll arrive at a certain location by a certain time because that's when their boss is expecting them.

All of these considerations help us think about what would really enhance our audience's lives in the medium term.

Awaken: Realizing Loftier Goals

If we can both attract users to our product or service by aligning it with what they think they need and solve a longer-term problem for them, ultimately they're going to stick around... if they feel like your product is matching their loftier goals. With the notion of *awakening*, I want you to think about soul searching. What would it mean to really awaken people's passions? That desire they've always had to learn to play the piano, or write a book, or complete the muddy adventure race?

Let's consider which of the Six Minds can help us awaken our audience's loftier goals:

Emotion

We want to think about setting our audience free to pursue these life goals. What would make them feel like they've "made it"? Having an amount of wealth that allows them flexibility to travel? Buying a big enough home to be able to host 5-course dinners for 12? Getting tenure as a professor? We want to understand what those drivers are for people. In solving their problem, we can point to how our solution gets them to their intended destination.

[SIDE NOTE]

Pinpointing underlying emotional drivers can also help establish a positive feedback loop with customers who are enjoying a product or service. Are there tangible ways that we've affected them positively? Throughout the life cycle of a product or service, as we show people the immediate and longer-term benefits and ultimately demonstrate how we are helping them with their big-picture goals, we're winning loyalty and brand ambassadors. Ideally, our clients start promoting our product for us because they like it so much. In my work, I've found there are ways to manage this type of life cycle. Because we're talking about people's deep-seated goals and corresponding emotions, this life cycle usually takes months. We want to think about what people are most hopeful for in the long term. What are they really trying to do, and what are they most afraid of that would stop them from getting there? What is the persona they're striving to become?

Memory

Part of answering the question of what customers want to become and what they might be fearful of that would stop them comes from memory. What constitutes success for them? Maybe for one person, everyone in their family was a farmer, and for them, the notion of success lies in breaking free of that mold and going to college. People often definine their goals by their past experience.

Decision Making

With problem solving, we want to get at what customers feel their long-term goal is. Part of that has to do with memory, but another part has to do with the steps they believe it takes to get to the point where they've "arrived." This involves defining their problem space and how they think they can move around in that space to achieve their goal.

These are all concepts we've talked about before, but here we're thinking about framing these insights into something that's practical for us as marketers or product designers. What would attract our audience to the product? What would keep them using it in the medium term? What would make them feel satisfied enough to keep using it, or even promote it?

CASE STUDY: BUILDERS

Challenge: One client we worked with sold construction products—think insulation, rebar, electrical wire, and the like. They had tools and technologies that performed better and were less expensive than some of the most popular products today. But what they were finding was that the builders, the ones doing the actual installing, were largely unwilling to change the way they were used to doing their job.

Our client was struggling with getting those who were set in their ways to adapt to new technologies that would help them in the long run. Here's how we used some of the Six Minds to help:

Problem solving

Through contextual interviews with the builders, we realized that they were laser-focused on efficiency more than anything else. Typically, they would offer the job at a fixed price, meaning that if any one project took longer than the time they estimated, they'd be losing money—and time they could be spending on other projects. The longer the job took, the less profit the builders were making, which gave them great incentive to complete jobs as efficiently as possible. In installing piping, for example, the builders would want to make sure—above cost or any other factor—that it was the fastest piping to install. So our client's high-performing, inexpensive, new-fangled piping actually presented a challenge to these builders because they would need to spend time training their staff on how to install it.

Attention

We haven't used attention as much in these examples, but this was one case where we found that getting the builders' attention was central to solving our client's problem. It was clear that the builders were only paying attention to reducing the time of projects. They weren't thinking about the long-term benefits of any one product over another, but rather the short-term implications of completing this project quickly so they could move on to the next project. Our client needed to market its products in such a way that they would be attractive to these busy, somewhat set-in-their-ways builders.

Language

> We observed that two very different languages were being spoken by the product manufacturer and the actual installers. The product manufacturers were using complex engineering terms, like "ProSeal Magnate," which actually made the installers feel even more uncertain of the new products because they weren't speaking the same language. In other words, they weren't picking up what the manufacturers were putting down. This uncertainty also evoked some mistrust, which we'll discuss momentarily.

Emotion

> We sensed some fear in these conversations. The builders were worried that the new material would not work as well, and they'd have to go back and reinstall it. Naturally, it made sense that they would stick to the familiar product they already knew how to install. In a broader sense, though, we saw that the builders' number one fear was losing the trust of the general contractor, who wields the power to give them more work on subsequent projects. Maintaining a trusting relationship with the general contractor was crucial to these installment contractors keeping their business going.

Outcome: To bring it all together, we considered what it was the builders were attending to, how they were trying to fix their perceived problem, the language they were using, and the emotional drivers at play. Our findings suggested a very different approach for our client, the product manufacturer. We recommended that they focus on the time-saving potential of the new materials rather than anything else. In branding and promoting the products, it would be essential to use language familiar to the builders, ensuring them that these could be installed faster and that they really worked. We recommended the manufacturer consider offering some free training and product samples to builders, and even reach out to general contractors to inform them of the benefits of the new products.

Some of these findings may sound more obvious and less like "aha moments," but that's often how we feel after analyzing our Six Minds data. It may not be earth-shattering, but if a contextual interview finding points to a consideration that we might have otherwise missed, especially through more traditional audience research channels, it can completely change how we design and sell our products.

Previously, the client hadn't considered any of these factors. Using the Six Minds, we were able to point to specific things like the fragile relationships between contractors and general contractors, and what these contractors were attending to through their language and emotions. Armed with these findings, we were able to make recommendations for a system that encourages sales around these cognitive and emotional drivers.

CASE STUDY: HIGH-NET-WORTH INDIVIDUALS

Challenge: In a very different example, another client in the financial industry wanted to explore what sorts of products or services they might be able to offer high-net-worth individuals. My team and I set out to try to discover some of the unmet needs this audience had. Here's what we found:

Attention

One thing we noted was not necessarily what our audience were focusing on, like with our builders, but what they were *not* focusing on. It's a gross understatement to say this group was busy. Whether they were young professionals, working parents, or retired adults, they filled their days to bursting with commitments and activities. They were in the office, they were seeing a personal trainer, they were picking up their kids from after-school care, they were cooking, they were doing community service, they were playing in a rec softball league. They were constantly racing to get as much done as possible and get the most out of life. Because of all the competing commitments, needs, and priorities tugging at them from different directions, our audience's attention was pretty scattered.

Emotion

It was clear that everyone in this audience had ambitions of productivity and success. Going further, however, we saw some key differences in the underlying goals of our audience depending on their life stage. The young professionals were making a lot of money, and many of them were just discovering themselves and starting to define what success and happiness meant to them. As you might imagine, the folks we talked to with small children had very different definitions of those concepts. They were focused on the success of the family unit. They wanted to make sure their kids had whatever they needed for everything from soccer practice to college.

Though members of this group were hyper-focused on family life, they were also worried about losing their sense of self. The older adults we talked to circled back to that first notion of self-discovery. One gentleman who wanted to keep exploring music built a bandstand in his basement so his friends could play with him. Another decided to follow his dream of taking historical tours; even though he knew it maybe wasn't "cool," it really made him happy.

Language

The differences in people's deep, underlying life goals also came through in the language they used. When we asked them to define "luxury," the young professionals mentioned first-class tickets and one-of-a-kind adventures in exotic places, getting at their deeper goals of self-discovery. The people with families talked about going out for dinner somewhere the kids could run around outside and they didn't have to worry about the dishes, getting at their goals of family togetherness, and also just maintaining sanity as a parent. Older adults like the gentleman I mentioned before talked about taking that trip of a lifetime, getting at their deeper goals of feeling like they've really lived and experienced everything they want to. As we know from considering language, something as simple as a word (e.g., "luxury") can have drastically different meanings to our different audiences.

Outcome: These findings, gleaned using the Six Minds, were the keys to making products specific to the needs of these different groups of high net worth individuals. When we offered our recommendations to the client, we focused on our biggest takeaway, which was that relative to the other populations, the older adults were really underserved.

When we looked at how credit cards and other banking instruments were being marketed, we found that they tended to target either young professionals (e.g., skydiving in Oahu) or families (e.g., 529 college savings plans). There was surprisingly little that targeted older adults around the financial tools they needed for self-discovery. Fortunately, we now had all this Six Minds data we were able to give our client to help change that.

Concrete Recommendations

- *Appeal* through a product's advertisements, promotions, and brand promise.

- Look at attention and vision. What is this audience looking for? What would draw them in? What is the language they use to describe what they're looking for?

- *Enhance* people's lives through the product or service design you are building.

- Look at decision making/problem solving, memory, and frameworks. What problem do your customers really need to solve, and what will ultimately solve that problem? How, if at all, will their framework and perspective have to change? What parts of their old metaphor work? Which parts do not?

- *Awaken* people's life goals.

- Look at deep emotions. What will resonate with this audience and link to their biggest goals and fears? What parts of the product can help to allay their fears? How can you help show progress toward their goals?

[17]

Succeed Fast, Succeed Often

WHILE I STILL ENCOURAGE industry-standard build–test–learn cycles, I do believe that the information gleaned from the Six Minds approach can get you to a successful solution faster if you start with this knowledge.

In this chapter we'll look at the Double Diamond approach to the design process, and I'll show you how you can use the Six Minds to narrow the range of possible options to explore and help you select the optimal designs. We'll also look at learning while making, prototyping, and contrasting new services or products with those of your competitors.

Up to now, we've focused on empathizing with our audience to understand the challenge from the perspective of the individuals who are experiencing it. The time we've taken to analyze the data should be well worth it, helping us to more clearly articulate the problem, identify the solution we want to focus on, and inform the design process to avoid waste.

I challenge the notion of "fail fast, fail often" because I believe the Six Minds approach can reduce the number of iteration cycles needed.

Divergent Thinking, Then Convergent Thinking

Many digital product and service designers are familiar with the Double Diamond product and service creation process, roughly summarized as consisting of four phrases: Discover, Define, Develop, Deliver (Figure 17-1).

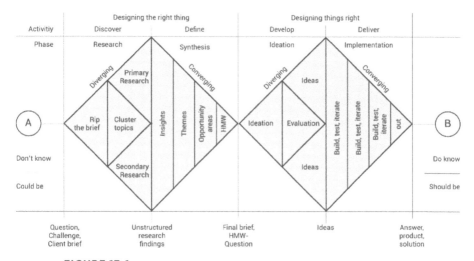

FIGURE 17-1

The Double Diamond design process

The Double Diamond process can appear complicated, so I want to make sure you understand how the Six Minds approach can help to focus your efforts along the way. I won't focus on discovery stages such as linking enterprise goals to user goals because there are wonderful books on the overall discovery process that have come out recently. I'll recommend them at the end of this chapter.

First Diamond: Discover and Define ("Designing the Right Thing")

Within a Double Diamond framework, we first have the Discover phase, where we try to empathize with the audience and understand what the problem is. Then comes the Define stage—figuring out which of those problems we might choose to focus on.

The Six Minds work generally would be seen as fitting into the discovery process. It provides a sophisticated but efficient way of capturing more information about the customers' cognitive processes and thinking, in addition to building empathy for their needs and problems. We are performing research revolving around understanding our customers better, creating insights, themes, and specific opportunity areas. Thinking back to the last chapter, we can use the results of our research to answer these questions:

- What *appeals* to our audience (what language they're using to describe what they want and what's drawing their attention)?

- What would *enhance* their lives (help them solve their problem and expand their framework of thinking about or interacting with our product or service)?

- What would *awaken* their passions (excite them and make them feel like they're accomplishing something important for themselves)?

Even though most of this chapter will focus on the second diamond, I want to acknowledge that our primary research is a treasure trove in terms of helping us identify specific opportunity areas. After considering the appeal/enhance/awaken triad, you'll probably have some insights into where opportunities exist, whether it's helping a wedding planner with finance management or helping a custom cabinet maker with marketing. Now you're ready to explore some of the possibilities for solutions.

Second Diamond: Develop and Deliver ("Designing Things Right")

By the time you get to the second diamond, you have selected a customer problem that you want to solve with your product or service. Now it's time to select the optimal design for the product or service that solves that problem.

There are a seemingly infinite number of solution routes you could take. To expedite the process of selecting the right one, you need ways to constrain the possible solution paths. The Six Minds framework is designed to dramatically reduce the product possibilities by informing your decision making. The answers to many of the questions posed in the analysis can inform design and reduce the need for design exploration. Here are a few examples:

Vision/Attention

Six Minds research can answer many questions for designers: What is the end audience looking for? What is attracting their attention? What types of words and images are they expecting? Where in the product or service might they be looking to get the information they desire? Given that understanding, the designers can determine how to use that knowledge to their advantage and whether to intentionally break from those expectations.

Wayfinding

> We will also have significant evidence for designing the interaction model, including answers to questions like the following: How does our audience expect to traverse the space, including virtually (e.g., walking through an airport or navigating a phone app)? What are the ways they're expecting to interact with the design (e.g., just clicking on something, or using a three-finger scroll, or pinching to zoom in)? What cues or breadcrumbs are they looking for to identify where they are (e.g., a hamburger icon to represent a restaurant, or different colors on a screen)? What interactions might be most helpful for them (e.g., double-tapping)?

Memory

> We will also have highly relevant information about user expectations: What past experiences are helping to shape their expectations for this one? What designs might be the most compatible or best synchronized with those expectations? What past examples will they be referencing as a basis for interacting with this new product? We can help to speed acceptance and build trust in what we are building by matching some of these expectations.

HAVE WE STIFLED INNOVATION?

"But what about innovation?!" you may be wondering. I am not saying "Thou shalt not innovate." There are certainly times when it's appropriate to come up with new kinds of interactions, new ways to draw attention, new paradigms. What I'm saying is that you should first consider whether there are any ways in which you can innovate within an existing body of knowledge and save yourself a huge amount of struggle. When we work within those existing precedents, we dramatically speed acceptance. Here are some things to keep in mind:

Language

> Content strategists will want to know to what extent customers are experts in an area. What language style would they find most comprehensible (e.g., would they say "the front of the brain" or "the anterior cingulate cortex")? What language would merit their trust and be the most useful to them?

Problem solving

What do the users believe the problem to be? In reality, is there more to the problem space than they realize? How might their expectations or beliefs have to change in order to actually solve the problem? For example, I might think I just need to get a passport, but it turns out that what I really need to do is first make sure I have documentation of my citizenship and identity before I can apply. How are users expecting us to make it clear where they are in the process?

Emotion

As we help people with their problem solving, how can we do this in a way that's consistent with their goals and even allays some of their fears? We want to first show them that we're helping with their short-term goals. Then we want to show them that what they're doing with our tool is consistent with those big-picture goals they have as well.

There are so many elements from the Six Minds that can help you as the designer to ideate more constructively, in a way that's consistent with the evidence you have—which means that your concepts are more likely to be successful when they get to the testing stage. Rather than just having a wide-open, sky's-the-limit field of ideas, you have all these clues about the direction your designs should take.

This also means you can spend more time on the overall concept, or on branding, versus debating some of the more basic interaction design.

Learning While Making: Design Thinking

Several times in this book, I've referenced the notion of design thinking. Popularized by the design studio IDEO, design thinking can be traced back to early ideas around how we formalize processes for creating industrial designs. The concept also has roots in some well-known psychological studies of systematic creativity and problem solving from the '70s by the psychologist and social scientist Herbert Simon, whose research on decision making I referenced earlier.

Think of building a product like a camera that will act as a surgeon's eyes during surgery. Obviously this is a tool you would need to design very carefully to ensure it could be manipulated precisely and bend the right ways. In working to construct a prototype, the engineers building this camera will learn a lot about things like the importance of its weight and grip. Similarly, there will be a myriad of things that you won't find out until you start making your product. That's why this ideation stage is literally thinking by designing.

Don't underestimate the importance of early sketches of the interactions and service flows. Bill Buxton, one of Microsoft's senior researchers, writes about this really well in his book *Sketching User Experiences: Getting the Design Right and the Right Design*. He proposes that any designer worth their salt should be able to come up with 7 to 10 ways of solving a problem in 10 minutes—not fully thought-out solutions, but quick sketches of different solutions and styles. Upon review, such sketches can help to inform which possible design directions might have merit and be worthy of further exploration.

Like Buxton, I think that really quick prototype sketching is very helpful and can show you just how divergent the solutions can be. When you review them, you can see the opportunities, challenges, and little gold nuggets in each one.

Starting with the Six Minds ensures that you approach this learning-while-making phase with some previously researched constraints and priorities in place. Rather than restraining you, these constraints will actually free you to build consensus on a possible solution space that will work best for your audience. You will be able to evaluate the prototypes through evidence-based decision making, focusing on what you've learned about the customer, rather than relying on the highest paid person's opinion (HIPPO).

Next, I'll give you one example of why it's so important to do your research prior to going ahead and building.

CASE STUDY: LET'S JUST BUILD IT!

I was working with a group on a design sprint. After I'd explained my process, the CEO said, "It's great that you have these processes, but we already know what we need to build." Generally, a team doesn't know what they need to build at this point, or if they're all really set on a particular direction, the reasons might not be good ones.

But the CEO wanted to jump in and get building, so we went right to the design to see what would happen.

As you see from the diagrams in Figure 17-2, their "solutions" were all over the map—the perception of the target audience and the problem to be solved varied widely between the team members. Quickly the CEO recognized that they were not aligned like he had thought. He graciously asked that we follow the systematic process after all.

I asked him and his team to quickly sketch out their solutions so I could see what it was they wanted to build (since they were on the same page and all).

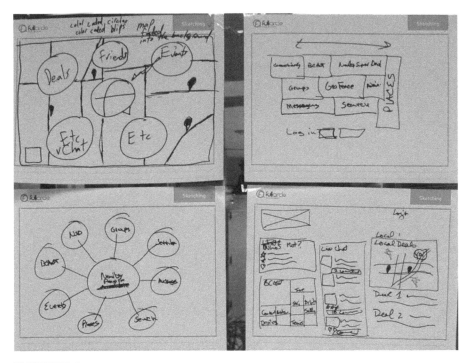

FIGURE 17-2
Vastly different visions for a website suggests the team members are not yet aligned

Don't Mind the Man Behind the Curtain: Prototype and Test

In the contextual inquiry, we consider where our target audience's eyes go, how they're interacting with our product or service, what words they're using, what past experiences they're referencing, what problems they think they're going to solve, and what some of their concerns or big goals are. I believe we can be more thoughtful in our build-test/learn cycle by taking these findings into account.

For the prototype stage, we'll go back to a lot of our Six Minds methods of contextual inquiry. We want to be watching where the eyes go with prototype 1 versus 2. We'll be looking at how customers seem to be interacting with our product, and what that tells us about their expectations. We'll consider the words they're using with these particular prototypes, and whether we're matching their level of expertise with the wording we're using. What are their expectations about using the prototype, or about the problem they need to solve? In what ways are we contradicting or confirming those expectations? What things are making them hesitate to interact with the product? (For example, are they not sure if the transaction is secure or if adding an item to the shopping cart means they've purchased it?)

In the prototyping process, we're coming full circle, and using what we learned from our empathy research early on. We're designing a prototype, or series of prototypes, to test all or part of our solution.

Here are a few observations and suggestions for you to keep in mind:

Avoid a prototype that is too high-fidelity

I've noticed that when we present a high-fidelity prototype—one that looks and feels as close as possible to the end product we have in mind—our audience starts to think that this is basically a fait accompli. It's already so shiny and polished and feels like it's actually live, even if it's not fully thought out yet. Something about the near-finished feel of these prototypes makes stakeholders assume it's too late to criticize. They might say it's pretty good, or that they wish this one little thing had been done differently, but in general it's good to go.

That's why I prefer to work with low-fidelity prototypes that still have a little bit of roughness around the edges. This makes the participants feel like they still have a say in the process and can influence the design and flow. Results vary with paper prototypes, so it's important that you know going into this exercise what the ideal level of specification is for this stage.

When I'm showing low- to medium-fidelity prototypes to a customer, for example, I prefer not to use the full branded color palette, but instead stick to black and white. I'll use a big X or hand sketch where an image would go. I'm cutting corners, but I'm doing that intentionally. Some of these rougher elements signal to the user that this is just an early concept that's still being worked out, and their input is still valuable in terms of framing how the end product should be built. I like to call this type of prototype a "Wizard of Oz" prototype—don't mind the man behind the curtain.

In one example of this, we were testing how we should design a search engine for a client. For this, we first wanted to understand the context of how people would be using the search engine. We didn't have a prototype to test, so we had them use an existing search engine. We used the example of needing to find a volleyball for someone who was eight or nine years old. We found that whether the users typed in "volleyball" or "kid's volleyball," the same search results came up. And that was OK because we weren't necessarily testing the precision of the search mechanism; we were testing how we should construct the search engine. We were testing how people interacted with the search, what types of results they were expecting, in what format/style, how they would want to filter those results, and generally how they would interact with the search tool. We were able to answer all of those questions without actually having a prototype to test.

Do in-situ prototyping

Now that you know I'm a fan of rough or low-fidelity prototypes, I also want to emphasize that you do still need to do your best to put people into the mode of thinking about what they will need. Going back to the contextual inquiry, I think it's important to do prototype testing at someone's actual place of work so they are thinking of real-world conditions.

Observe, observe, observe

This is the full-circle part. When we're testing the prototype, we observe the Six Minds just like we did in our initial research process. Where are the users' eyes going? How are they attempting to interact? What words are they using at this moment? What experiences are they using to frame this experience? How are we being consistent with or breaking those anticipations? Do they feel like they're actually solving the problem that they have? Going one step further, a more subtle question: how does this show them that their initial concept of the problem may have been impoverished, and that they're better off now that they're using this prototype?

When we think of emotion here, it's not very typical for an early prototype to show someone that they're accomplishing their deepest life goals. But we can learn a lot through people's fears at this stage. If someone has deep-seated fears, like those young ad execs buying millions of dollars in ads, we can observe through the prototype what's stopping the users from acting, or where they're hesitating, or what seems unclear to them.

Test with Competitors/Comparables

As you're doing these early prototype tests, I strongly encourage you test them with live competitors if you can. One example of this was when we tested a way that academics could search for papers. In this case, we had a clickable prototype so that people could type in something, even though the search function didn't work yet. We tested it against a Google search and another academic publishing search engine. Like with the volleyball example earlier, we wanted to test things like how we should display the search results and how we should create the search interface, in contrast with how these competitors were doing those things.

With this sort of comparable testing before you've built your product, you can learn a lot about new avenues to explore that will put you ahead of your competitors. Don't be afraid to do this even if your tool is still in development. Don't be afraid of crashing and burning in contrast to your competitors' slickest products—or in contrast to your existing products.

I also recommend presenting several of your own prototypes. I'm pretty sure that every time we've shown a single prototype to users, their reaction has been positive: "It's pretty good," "I like it," "Good work." When we contrast three prototypes, however, they provide much more substantive feedback. They're able to articulate which parts of prototype 1 they just can't stand, versus the aspects of prototype 2 that they really like, if we could possibly combine them with this component of prototype 3.

There's also plenty of literature backing up this approach. Comparison reveals further unmet needs or nuances in interfaces that the interviews might not have brought out, or beneficial features that none of the existing options are offering.

Concrete Recommendations

- Simulate the product and test your design direction with users (including simulating AI systems).

- Use the same methodologies described here to further test your understanding of users' cognitive experience (e.g., vision/attention, wayfinding, language, memory/assumptions, decision making, and emotion).

- Rework failures to be more consistent with underlying cognitive systems as a way to reduce the number of failures as you build and explore the solution space.

Further Reading

Buxton, B. (2007). *Sketching User Experiences: Getting the Design Right and the Right Design*. San Fransisco: Morgan Kaufmann.

[18]

Now See What You've Done?

CONGRATULATIONS! YOU ARE READY to design an experience based on multiple levels of human experience, and can test the experience of your product or service more systematically than ever before by using the Six Minds framework. Be prepared to get to better designs faster, and with less debate over design direction.

In this chapter, I'll present a summary of everything we've looked at up to this point. I'll also provide a few examples of some of the kinds of outcomes you might expect when designing using the Six Minds.

One of the things that I think is unique to this approach is the notion of empathy on multiple levels. Not only are we empathizing with the problem our audience is trying to solve, but we're also taking into account several other cognitive systems when making design decisions. By focusing on specific aspects of the experience (e.g., language, or decision making, or emotional qualities), our Six Minds approach allows us to undertake the decision-making process with much more evidence than if we had relied on more traditional audience research channels.

The last thing I want to speak to in this chapter goes back to something I mentioned back in Chapter 1, about all the elements that together make an experience brilliant—and when I say "experience," I'm actually thinking of the series of little experiences that add up to what we think of as a singular experience. The experience of going to the airport, for example, is made up of many small experiences, like being dropped off at the curbside, finding a kiosk to print your boarding pass, checking your bag, going through security, getting to customs, finding the right terminal, heading to your gate, buying a snack, etc. In many cases, our "experience" actually involves a string of experiences, not a singular moment in time, and I think we need to keep this realization in mind as we design.

Empathy on Multiple Levels

In Lean Startup speak, people talk about GOOB, which stands for Get Out Of (the) Building. In traditional design thinking, empathy research starts with simply seeing the context in which your actual users live, work, and play. We need to empathize with our target audience to understand what their needs and issues really are. There are some great people out there who can do this by intuition. But for the rest of us mortals, I think there are ways to systematize this type of research. If you do it the way I proposed in Part II of this book, through contextual inquiry, you'll come home with pages of notes, scribbles, sketches, diagrams, and interview tapes.

Findings from each of the Six Minds won't necessarily influence every design decision. But below, I provide a representative example where I think they all come into play (Figures 18-1 and 18-2).

FIGURE 18-1

How an understanding of the customer's vision, wayfinding, and memory needs influenced the design of the PayPal for business website

FIGURE 18-2

How an understanding of the customer's emotions, language, and problem solving goals directly affected the design of the PayPal for business website

In this example, we were designing a page for PayPal for small businesses. The end users were people who might consider using PayPal for their small business ventures, allowing them to receive credit card payments on their websites or accept in-person credit card payments instead of cash. Let's put our design to the Six Minds test, based on the interviews we conducted and watching people work:

Vision/Attention

We designed this page to only have one image at the top. It's darker than the rest of the page and has much more visual complexity. Inevitably, people's attention will be drawn to that picture. On top of it, we placed white type that's much bigger than the rest, and stands out against the dark image. People's eyes are drawn to that box of text. The other visual thing is that we wanted to make it clear how small business owners could sign up for PayPal. That's why we made sure our sign-up button visually "pops out" in terms of its blue color and shape relative to the background image.

Language

The text at the top of the page simply says "Businesses sell more with PayPal." It's very straightforward language. It doesn't use a lot of fancy marketing terms. It matches word-for-word what we heard business owners say they wanted to accomplish. It literally speaks their language. This was intentional. We found that the majority of the business owners we talked to were quite new to ecommerce and credit card processing. They wanted to add PayPal because they didn't want to slow down anyone making purchases through their websites. That's why we added buttons that say things like "add PayPal to your checkout," and "offer credit cards + PayPal"—these statements were taken verbatim from our interviews. Our interviewees told us they wanted the presentation to be straightforward, so why complicate things by putting a marketing spin on their words?

Memory

We wanted to appeal to people's memory and interpretation of the picture to set the tone of what we're talking about here. The image looks like it's in the back of a coffee shop. You can see big sacks of coffee beans and it looks like these two casually dressed men are looking at some of the beans. The image doesn't give off a big, corporate vibe. It feels more like a small, two-man, maybe even family-owned operation, a sort of boutique coffee shop where the owners would know the names of their "regulars." We're trying to evoke something smaller and tucked away, the work of two artisans who really understand their craft. Just by using one image, we can set the stage and let small business owners know they're in the right place.

Wayfinding

In light gray, above the fold, we used some navigation to show users what's available further down the page: "Learn about us," Website Payments, Point-of-Sale Payments, and Online Invoicing. This navigation bar shows the audience both where they are and where they can jump to next. It orients them to what this page is all about and how they can interact with it. We also introduced an equivalent navigation in the responsive mobile version.

Decision Making

At the end of the day, we wanted to awaken our audience's passion while at the same time providing rational reasons for acting. We knew the problem these business owners were trying to solve: how to sell more. At the bottom of the page, we added four statistics about companies that use PayPal for their business and what selling more has looked like for them. We provided these statistics: because business owners and partners need the cold, hard, logical rationale before they're going to jump into any business decision with both feet.

Emotion

We also wanted to get people excited about the possibility of selling more. Tapping into the audience's immediate objective of selling more (appeal), we stated—twice—in our mock-up that businesses sell more with PayPal. The statement reinforces that our audience could be selling more, which opens up endless possibilities for their lives (enhance) and even their overall sense of success and identity. We wanted to awaken this longing to reinforce the possibility of their success as business owners.

Evidence-Driven Decision Making

In the example we just looked at, just in this portion of a webpage, we talked about all Six Minds and how we can use evidence-driven decision making when we're formulating a product design or determining what direction to go in. I believe, of course, that this process gives us much clearer input than traditional types of prototyping and user testing.

That said, getting to this mock-up didn't happen overnight, or even halfway through our contextual interviews. Sure, we saw some patterns and inklings through our Six Minds analysis. But getting to that actual design was a slow and gradual process. We tried out many iterations and made microdecisions as we went based on customer input. We also considered comparable sites and some of the weaknesses that we found in those to make sure we did better than all of them.

I believe there is much we can learn through the process of just formulating these designs, or "design thinking."

Figure 18-3 shows some early sketches of the different ideas we had for what the page would look like, including things like flow, functionality, and visuals. We started with lots of sketches and possibilities, doing a lot of ideating, rapid prototyping, and considering alternatives. As we narrowed down the possibilities through user testing and our own observations, we went from our really simple sketches to black-and-white mock-ups to clickable prototypes to the very high-fidelity one that you saw a little earlier in the chapter.

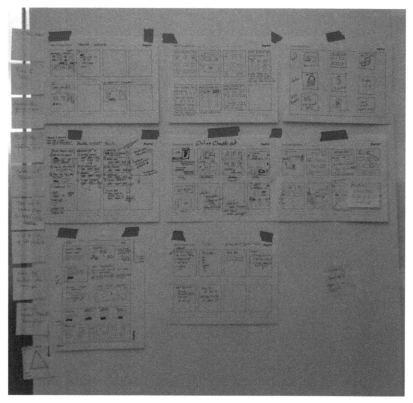

FIGURE 18-3
Design brainstorming: identifying strong concepts consistent with customer needs and iterating

Experience Over Time

The example we just walked through was a snapshot of someone's decision to sign up for PayPal for Business at a certain point in time. I want to go one step further here and show you how the Six Minds are applicable throughout the life cycle of a decision, and can actually be quite fluid, rather than static, over time.

Service design is a good example of the lifecycle of a decision. In this case, Figure 18-4 shows sticky notes with all the questions we heard from business owners about why they would or wouldn't consider PayPal for Business to start an ecommerce store.

FIGURE 18-4
Creating a journey map through all the microquestions customers have before they commit to a purchase

When we looked at all the questions together, we saw that they extended from a pretty fundamental level (e.g., "Can this do what I need?"), to follow-up questions (e.g., "Is the price fair?"), to implementation concerns (e.g., "Is this compatible with my website provider?"), all the way to emotions like fear (e.g., "What happens to me if someone hacks the system?"). We organized these questions into several key steps along the decision-making continuum.

People's questions, concerns, and objections tend to get more and more specific as they go. As you test your system, take note of when different questions come up in the process. Then you can design a system that presents information and answers questions at the logical time. It may be that in the steps right before they hit "buy," you're presenting things to customers in a more sophisticated way because they now already know the basics of what you're offering in relation to what they're currently using. Now you can answer those last questions that might relate to any fears that are holding them back.

Multiple Vantage Points

In summary, I first encourage you to embrace the notion of user experience as multidimensional and multisensory. We can and should tap into these multiple dimensions and levels when we're doing empathy research and design.

Second, as you encounter microdecisions within your product or service, remember that even these seemingly small steps can have a logical rationale based on your interviews. Embrace this rationale, especially in the face of HIPPO opposition—and embrace your creativity while you're at it. The type of evidence-driven design methodology I'm suggesting allows for plenty of creativity within what's more likely to be a winning sphere.

Last, try thinking about your product or service as more than just a transaction. Think about it as a process that will touch multiple people, over multiple points in time. Try thinking of the Six Minds of your product—where people's attention is, how they interact with it, how they expect this experience to go, the words they are using to describe it, what problem they're trying to solve, and what's really driving them—as constantly evolving. The more they learn about this topic through your product or service, the more expert they'll become, which changes how you engage them, the language you use, and so on.

Concrete Recommendations

- Consider your users' experience over time:

    - How will their behaviors change over time as they gain expertise with this product and in this area?

    - How will their problem space change over time?

    - How will their language and semantic representations of the words change?

[19]

How to Make a Better Human

GROWING UP, I REMEMBER watching reruns of a 1970s TV show about a NASA pilot who had a terrible crash, but according to the plot, scientists said, "We can rebuild him; we have the technology" and the pilot went on to become "the Six Million Dollar Man" (now the equivalent of about $40 million). With one eye with exceptional zoom vision, and one arm and two legs that were bionic, he could run 60 miles per hour and perform great feats. In the show he used his superhuman strength as a spy for good.

It was one of the first mainstream shows that demonstrated what might be accomplished by seamlessly bringing together technology with humans. I'm not sure that it would have been as commercially viable if they had called the show *The Six Million Dollar Cyborg*, but that's really what he was—a human–machine combination.

Today we have new possibilities with the resurgence and hype surrounding AI and ML. I'm sure if you are in the product management, product design, or innovation space, you've heard all kinds of prognostications about what might be possible. Here, I'd like to suggest what might be the most powerful combination: thinking about human computational styles and supporting them with ML-equipped experiences to create not the physical prowess of the Six Million Dollar Man, but the mental prowess that they never explored in the TV show.

Symbolic AI and the AI Winter

I'm not sure if you are aware, but as I write this we are in at least the second cycle of hype and promise surrounding AI. In the 1950s and 1960s, Alan Turing posited that mathematically, 0s and 1s could represent any type of mathematical deduction, suggesting that computers could perform formal reasoning. From there, scientists in neurobiology and information processing started to wonder, given the similarity of brain neurons' ability to fire an action potential or not (effectively a one or zero), if there might be the possibility of creating an artificial brain, and capable of reasoning. Turing proposed the Turing test: essentially, if you could pass questions through to some entity, and that entity could provide answers back, and humans couldn't distinguish between the artificial system's responses and that of a real human, then that system passed,and could be considered AI.

From there, others, like Herbert Simon, Allen Newell, and Marvin Minsky, started to look at intelligent behavior that could be formally represented and work on how "expert systems" could be built up with an understanding of the world. Their artificial intelligence machines tackled some basic language tasks, games like checkers, and some analogical reasoning. There were bold predictions that within a generation the problem of AI would largely be solved.

Unfortunately, their approach showed promise in some fields, but great limits in others—in part because it focused on symbolic processing, very high level reasoning, logic, and problem solving. This symbolic approach to thinking did find success in other areas, including semantics, language, and cognitive science, but the focus was much more on understanding human intelligence than building generalized AI.

By the 1970s, money for AI in the academic world had dried up, and there was what was called the "AI Winter"—going from the amazing promise of the 1950s to real limitations in the 1970s.

Artificial Neural Networks and Statistical Learning

Very different approaches to AI and the notion of creating an "artificial brain" started to be considered in the 1970s and 1980s. Scientists in a divergent set of fields composing cognitive science (psychology, linguistics, computer science), particularly David Rumelhart and James McClelland, looked at this from a very different, "subsymbolic" approach. Perhaps rather than trying to build representations that were used by humans, they hypothesized, we could instead build systems like brains—systems that had many individual processes (like neurons) that could affect one another with inhibition or excitation (like neurons) and have "back-propagation" that changed the connections between the artificial neurons depending on whether the output of the system was correct.

This approach was radically different because: (a) it was a much more "brain-like" parallel distributed processing (PDP), in comparison to a series of computer commands; (b) it focused much more on statistical learning; and (c) the programmers didn't explicitly provide the information structure, but rather sought to have the PDP learn through trial and error and adjust the weights between its artificial neurons itself.

These PDP models had interesting successes in natural language processing and perception. Unlike the symbolic efforts in the first wave, this group did not make any assumptions about how these ML systems would represent the information. These systems are the underpinnings of Google TensorFlow and Facebook Torch. It is this type of parallel process that is responsible for today's self-driving cars and voice interfaces.

With the incredible resources available in mobile phones and the cloud, modern systems have the computing power Newell and Simon likely never even dreamed of having. But while great strides have been made in natural language processing and image processing, these systems are still far from perfect, as shown in Figure 19-1.

There have been many breathless prognostications about the power of AI and its unstoppable intelligence. While these systems have been getting better, they are highly dependent on having the data available to train them and still have their limitations.

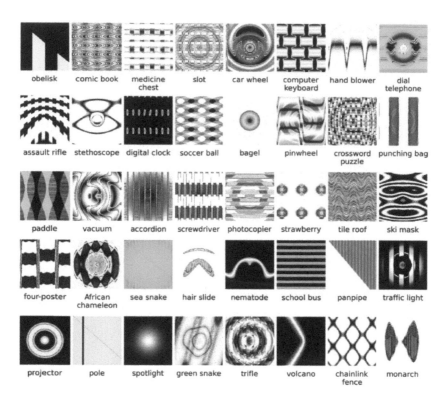

FIGURE 19-1

Less-than-perfect captions assigned by an ML algorithm

I Didn't Say That, Siri!

You may have your own experiences with how voice commands can on the one hand be incredibly powerful, but on the other have significant limitations. Their ability to recognize any language at all is impressive. This is a hard problem and they have shown real ability to do solve it. We put these systems to the test, studying Apple Siri, Google Assistant, Amazon Alexa, Microsoft Cortana, and Hound. Using a *Jeopardy!*-like setup, we asked participants to create a command or question using the provided terms, designed to get an answer (e.g., "Cincinnati, tomorrow, weather," for which participants might say, "Hey Siri, what is the weather tomorrow in Cincinnati?").

To make a long story short, we found that these systems were quite good at answering questions about basic facts (e.g., the weather, or the capital of a country), but had real trouble with two very natural human abilities. First, humans can put together ideas easily (e.g., population, country with Eiffel Tower—which we know is France). When we asked these systems a question like "What is the population of the country with the Eiffel tower?" they generally produced the population of Paris or just gave an error. Second, we can guage context. If we asked "What is the weather in Cincinnati?" and followed up with, "How about the next day?" these systems were generally unable to follow the thread of the conversation.

In addition, we found the humans experiencing these systems had a significant preference for the AI systems that responded in the most humanistic way—even if that system got something incorrect or was unable to answer (e.g., "I don't know how to answer that yet"). When it addressed the participants in the way they addressed it, they were most satisfied.

But is Siri really smart? Intelligent? It can add a reminder and turn on music, but you can't ask it if it is a good idea to purchase a certain car, or how to get out of an escape room. It has limited, ML-based answers. It is not "intelligent" in a way that would pass the Turing test.

The Six Minds and AI

Interestingly, the first wave of AI was known for its strength in performing analogies and reasoning (memory, decision making, and problem solving), and the more recent approach has been much more successful with voice and image recognition (vision, attention, language). The systems that provided more human-like responses tend to be favored (emotion).

I hope you are seeing where I am headed. The current systems are starting to show the limitations of a brute force, purely statistical, subsymbolic representation. While these systems are without a doubt amazingly powerful and fantastic for solving certain problems, no amount of faster chips or new training regimens will achieve the goals of AI sought out in the 1950s.

If more speed isn't the answer, what is? Some of the most prominent scientists in the ML and AI fields are suggesting we take another look at the human mind. If studying the individual—and group neuron level achieved this success in the perceptual realm, perhaps considering other levels of representation will provide even more success at the symbolic level with vision/attention, wayfinding and representations of space, language and semantics, memory, and decision making.

Just like with traditional product and service design, you might expect that I would encourage those building AI systems to consider the representations you're using as inputs and outputs, and test representations that are at different symbolic levels (e.g., word-level, semantic-level), rather than purely perceptual levels (e.g., pixels, phonemes, sounds).

I Get By with a Little Help from My (AI) Friends

While AI and ML researchers seek to produce independently intelligent systems, it is very likely that more near-term successes can be achieved using AI and ML tools as cognitive support tools. We already have many of these right now on our mobile devices. We can remember things using reminders, translate street signs with our smartphones, get help with directions from mapping programs, and get encouragement to achieve goals with programs that count calories and help us save money, or get more sleep or exercise.

In our studies of voice-activated systems today, however, the biggest challenges we've seen are the differences between the language employed by users versus that used by the system, and when the assistance was provided relative to when it was needed. When building things that allow customers or workers to do things faster and more easily by augmenting their cognitive abilities, the Six Minds can be an excellent framing of how ML and AI can support human endeavors:

Vision/Attention

AI tools, particularly with cameras, could easily help to draw attention to the important parts of a scene. They could help bring relevant information into focus (e.g., what form elements are unfinished), or if they know what you are seeking, highlight relevant words on a page or in parts of a scene. Any number of possibilities come to mind. When entering a hotel room for the first time, people want to know where the light switches are, how to change the temperature, and where the outlets are to recharge their devices. Imagine looking through your glasses and having these things highlighted in your view.

Wayfinding

Given successes with Lidar and automated cars, it seems likely that the type of heads-up display I just mentioned could also bring into attention the highway exit you need to choose, that tucked away subway entrance, or the store you might be seeking in the mall. Much like game playing, it could show two views—the immediate scene in front of you, and a bird's-eye map of the area and where you are in that space.

Memory/Language

We work with a number of major retailers and financial institutions who seek to provide personalization in their digital offerings. By getting evidence through search terms, clickstreams, communications, and surveys, one could easily see the organization and the terminology of the system being tailored to the individual. Video is a good example, where some customers might be just starting out and need a good camera for YouTube videos, while others might be seeking specific types of ENG (electronic news-gathering) cameras with 4:2:2 color, etc. Neither group really wants to see the other's offerings in their search, and the language and detail that each group needs would be very different.

Decision Making

I have discussed the fact that problem solving is really a process of breaking down large problems into their component parts and solving each of these subproblems. In each step, you have to make decisions about your next move. Buying a printer is a good example. A design studio might want a larger-format printer with very accurate colors. A law firm might want legal paper handling, good multiuser functionality, and the ability to bill the printing back to the client automatically. A parent with school-aged kids might want a quick, durable color printer all family members can use. By asking a little about the needs of the individual, and supporting each of the microdecisions that need to be made along the way (e.g., What is the price? How much is toner? Can it print on different sizes of paper? Do I need double-sided printing? What reviews are there from families?), the ML/AI might be able to intuit the types of goals the individual might have. The individual's location in the problem space might suggest exactly what that person should and shouldn't be presented with at this time.

Perhaps one of the most interesting possibilities is that increasingly accurate systems for detecting facial expressions, movement and speech patterns can ascertain the user's emotional state, which could be used to moderate the amount presented on a screen, the words used (perhaps the user is overwhelmed and wants a simpler route to an answer).

Endless possibilities abound, but they all revolve around what the individual is trying to accomplish, how they think they can accomplish it, what they are looking for right now, the words they expect, how they believe they can interact with the system, and where they are looking. I hope that framing your problem in terms of the Six Minds will allow you and your team to exceed all previous attempts at satisfying your users with a brilliant experience. I hope you can heighten every one of your users' cognitive processes in reality, just as that fictional team of scientists augmented the physical capabilities of the Six Million Dollar Man.

Concrete Recommendations

- Suggest different ways of training AI systems explicitly for semantics (rather than skipping this).

- Consider explicitly training AI systems in specific types of syntactic patterns that were less common in the findings you collected.

- Think about how you want to augment cognition (directing attention, encouraging certain kinds of interactions, providing information persuasively, etc.).

[*Appendix*]

Recommended Reading

Part I

Ariely, D. (2008). *Predictably Irrational: The Hidden Forces that Shape Our Decisions.* New York: HarperCollins.

Brafman, O., & Brafman, R. (2008). *Sway: The Irresistible Pull of Irrational Behavior.* New York: Crown Business.

Cialdini, R. B. (2006). *Influence: The Psychology of Persuasion, Revised Edition.* New York: Harper Business.

Evans, J. S. B. T. (2008). "Dual-Processing Accounts of Reasoning, Judgment, and Social Cognition." *Annual Review of Psychology* 59: 255–278.

Evans, J. S. B. T., & Stanovich, K. E. (2013). "Dual-Process Theories of Higher Cognition: Advancing the Debate." *Perspectives on Psychological Science* 8(3): 223–241.

Gallistel, C. R. (1990). *The Organization of Learning.* Cambridge, MA: MIT Press.

Gladwell, M. (2014). *Blink: The Power of Thinking Without Thinking.* New York: Back Bay Books.

Intraub, H., & Richardson, M. (1989). "Wide-Angle Memories of Close-Up Scenes." *Journal of Experimental Psychology: Learning, Memory, and Cognition.* *https://doi.org/10.1037/0278-7393.15.2.179*

Kahneman, D. (2011). *Thinking Fast and Slow.* New York: Macmillan.

LeDoux, J.E. (1996). *The Emotional Brain: The Mysterious Underpinnings of Emotional Life.* New York: Simon & Schuster.

Müller, M., & Wehner, R. (1988). "Path Integration in Desert Ants, Cataglyphis Fortis." *Proceedings of the National Academy of Sciences* 85(14): 5287–5290.

Pink, D. H. (2009). *Drive The Surprising Truth About What Motivates Us.* New York: Riverhead Books.

Power, M., & Dalgleish, T. (1997). *Cognition and Emotion: From Order to Disorder.* Hove, Englad: Psychology Press.

Simon, H. A. (1956). "Rational Choice and the Structure of the Environment." *Psychological Review* 63(2): 129–138.

Thaler, R., & Sunstein, C. (2008). *Nudge: Improving Decisions About Health, Wealth and Happiness.* New York: Penguin Books.

Tversky, A., & Kahneman, D. (1981). "The Framing of Decisions and the Psychology of Choice." *Science* 211(4481): 453–458.

Tversky, A., & Kahneman, D. (1974). "Judgment Under Uncertainty: Heuristics and Biases." *Science* 185(4157): 1124–1131.

Wong, K., Wadee, F., Ellenblum, G., & McCloskey, M. (2018). "The Devil's in the g-Tails: Deficient Letter-Shape Knowledge and Awareness Despite Massive VisualExperience." *Journal of Experimental Psychology: Human Perception and Performance.* 44(9): 1324–1335. *https://doi. org/10.1037/xhp0000532*

Part II

Chipchase, J. (2007). "The Anthropology of Mobile Phones" TED Talk. Retrieved January 15, 2019, from *http://bit.ly/2Uy9J1A.*

Chipchase, J., Lee, P., & Maurer, B. (2011). Mobile Money: Afghanistan. *Innovations: Technology, Governance, Globalization.* 6(2): 13–33.

IDEO.org. (2015). "The Field Guide to Human-Centered Design." Retrieved January 15, 2019, from *http://www.designkit.org// resources/1.*

Part III

Buxton, B. (2007). *Sketching User Experiences: Getting the Design Right and the Right Design.* San Fransisco: Morgan Kaufmann.

[*Index*]

A

abstract concepts, 31–37
acuity, visual, 16–17
affinities and psychological profiles, 141–143
AI Winter, 196
Alexa (Amazon), 28–29, 198
Amazon Alexa, 28–29, 198
analyzing dreams, 55, 59–60, 131–133
analyzing user research. *See* Post-It note categorization method
analyzing word frequency, 96–97
appeal (form of emotion), 130–131, 161–163, 175
Apple Siri, 28, 198–199
Ariely, Dan, 57
artifacts (what researchers notice), 68
artificial intelligence
    artificial neural networks, 197–199
    background information, 196–197
    recommendations for, 202
    statistical learning, 197–199
Assistant app (Google), 198
assumptions
    challenging internal, 155–156
    contextual interviews and, 71, 81
    empathy research and, 65–66
attention. *See* vision, attention, and automaticity
audience segmentation
    case study, Millennial money, 152–154
    case study, trust in credit, 154
    challenging internal assumptions, 155–156
    creating, 77–78, 113, 141, 152–154
    emotion and, 147–148
    empathy research and, 155–158

finding the dimensions, 152–155
identifying, 117
language and, 143–146
psychographic profiles, 131–132, 141–143
recommendations for, 160
wayfinding and, 149–151
augmented reality (AR), 24
automaticity. *See* vision, attention, and automaticity
awaken (form of emotion), 130–131, 165–166, 175

B

behaviors, unconscious, 12–14, 64, 70
blockers (problems), 52–53
boundary extension, 35–36
brain
    artificial intelligence and, 196–198
    spatial information and, 19–21
    too much information, 56–57
    what information/pathway, 9–10, 69–70
    where information/pathway, 9–10, 21–22, 28–29
Buxton, Bill, 178

C

Cancer.gov website, 45
case studies
    adventure race, 133–135
    auction website, 90–91
    builders, 167–169
    coupons, 127
    credit card theft, 131
    distracted movie watching, 109–110
    ecommerce payment, 122–123

focus groups, 63
framing problems differently, 49–52
F-shaped eye search pattern, 13–14

G

Gallistel, Randy, 19
goals (customer), 55, 59–60, 122,
    131–133
goal state (problem solving), 48, 122
GOOB (Get Out Of Building), 186
Google products
    Assistant app, 198
    Home assistant, 28
    TensorFlow software library, 197

H

"happy hour" concept, 37
heat maps, 13, 84, 88–90
high-fidelity prototypes, 180–181,
    190
Home assistant (Google), 28
Hound app (SoundHound), 198
Human-Centered Design Toolkit
    (IDEO), 65

I

icons, 16
IDEO design studio, 65, 177
innovation, 176–177
in-person investigations. *See* contextual interviews
in-situ prototyping, 181
Instagram social media platform, 16
interfaces
    observing user interactions, 25–
        28, 104–106
    voice-activated, 28–29, 198–200
interruptions (what researchers
    notice), 68
interviews. *See* contextual interviews
"in the moment" and memory, 64, 81
irrationality in decision making, 57–
    58

J

journey map, 127, 191

K

Kahneman, Daniel, 11, 47, 56
Kaplan, Craig, 50

L

language
    about, 4–5, 7–8, 78
    appeal form of emotion and, 162
    audience segmentation and, 143–
        146
    case study, builders, 168
    case study, high-net-worth individuals, 170
    case study, Institute of Museum
        and Library Services, 101
    case study, medical terms, 97–98
    communication difficulties, 42–
        44, 49
    Double Diamond process
        and, 176
    empathy in design decisions, 188
    machine learning/artificial intelligence and, 201
    Post-It note categorization
        method, 75, 79–81, 98–101,
        143–146, 153
    questions regarding customers, 61
    reading between the lines, 96–98
    recommendations for, 71, 102
    recording interviews, 96
    revealing level of expertise, 44–45,
        96–98, 143–146
    See/Feel/Say/Do chart and, 158–
        159
    semantic concepts and, 41
    uncovering semantic representations in interviews, 46
    word frequency analysis, 96–97
Lean Startup methodology, 186
learning-while-making phase, 177–
    179
LeDoux, Joseph, 57
LifeHacker website, 163

M

machine learning (ML), 200–202
market research. *See* user research
McClelland, James, 197
MedlinePlus website, 97–98

memory
about, 5, 7–8, 78
abstract concepts, 31–34
awaken form of emotion and, 166
boundary extension, 35–36
case study, producing products vs. managing business, 112
case study, tax code, 113
case study, timeline of researcher's story, 118–119
context triggering, 64, 81
Double Diamond process and, 176
empathy in design decisions, 188
enhance form of emotion and, 164–165
expectations and, 5, 36–37, 39, 106–107
machine learning/artificial intelligence and, 201
meanings in the mind, 112–113
Post-It note categorization method, 75, 79–81, 114–116, 154
questions regarding customers, 61, 111
recommendations for, 71, 119
See/Feel/Say/Do chart and, 159–160
stereotypes, 31–32, 36–37
trash talk experiment, 33–36
understanding mental models, 38–39
mental models, 38–39
Microsoft Cortana, 28, 198
Minsky, Marvin, 196
mock-ups, 189–190
motion, visual popout and, 14
Mural tool, 75
mutilated checkerboard problem, 50–52

N

navigational cues
observing user interactions, 25–28, 104–106
physical vs. virtual space, 24–25
voice interfaces and, 28–29
neural networks, artificial, 197–199
Newell, Alan, 196–197

novices
language and, 44–45, 96–98, 143–146
problem solving and, 49–53
null result in eye tracking, 15

O

The Organization of Learning (Gallistel), 19

P

parallel distributed process (PDP), 197
PayPal (company), 186–191
physical space, wayfinding in, 21–29, 104–107
Pinterest social media platform, 24
popout, visual, 14–15
Post-It note categorization method
affinities and psychological profiles, 141–143
analysis exercise, 78–82
creating audience segmentation, 77–78, 152–154
decision making category, 75, 79–81, 124–127, 153
emotion category, 76, 79–81, 136–138, 147–148, 153
language category, 75, 79–81, 98–101, 143–146, 153
looking for trends across participants, 77–78
memory category, 75, 79–81, 114–116, 154
organizing participant findings, 76–77
reviewing and writing down observations, 75–76
vision category, 75, 79–81, 91–94
wayfinding category, 75, 79–81, 107–109, 149–151, 154
Predictably Irrational (Ariely), 57
problem solving
building subgoals, 48–49, 52–53, 122
case study, builders, 167
defining problem, 48–49
Double Diamond process and, 177

framing problems differently, 49–52

machine learning/artificial intelligence and, 201–202

matching audience perception of virtual space, 28

mutilated checkerboard problem, 50–52

resolving problems, 52

prototyping, 178–183, 190

psychographic profiles, 131–132, 141–143

Q

questions regarding customers
contextual interviews, 74–75
decision making, 61, 121
emotion, 62, 129
language, 61, 95
memory, 61, 111
questioning existing assumptions, 71, 81
vision, attention, and automaticity, 61, 83–84, 90
wayfinding, 61, 103

R

RealTimeBoard tool, 75
redefining the problem space, 49–52
risk aversity, 57
Rumelhart, David, 197

S

saccades. See eye movement and tracking
satisficing, 55–56, 58–59, 135–136
See/Feel/Say/Do chart, 156–159
semantics. See also memory
about, 41
considering customer associations, 111–119
mental models and, 38–39
semantic map, 42
stereotypes and, 36
uncovering representations in interviews, 46
Simon, Herbert, 50, 56, 177, 196–197
Siri (Apple), 28, 198–199

Six Million Dollar Man (TV show), 195
Six Minds of Experience
about, 6–7, 78
affinities and psychological profiles, 141–143
data-to-insights process. See Post-It note categorization method
decision making. See decision making
Double Diamond process and, 173–177
emotion. See emotion
finding the dimensions, 152–155
language. See language
learning-while-making phase, 177–179
memory. See memory
recommendations, 70–74, 183
user research considerations, 61–62
vision, attention, and automaticity. See vision, attention, and automaticity
wayfinding. See wayfinding
Sketching User Experiences (Buxton), 178
slow thinking (conscious processes), 11–12, 56
Snapchat social media platform, 23
sophistication of language usage, 96–98
"stalking with permission". See contextual interviews
statistical learning, 197–199
stereotypes, 31–32, 36–37, 112–113
sticky notes. See Post-It note categorization method
subgoals in problem solving, 48–49, 52–53, 122
"succeed fast, succeed often" approach
about, 173
design thinking and, 177–179
Double Diamond process, 173–177
prototyping and testing, 179–183
recommendations, 183
superusers, 64

surveys, 63, 69–70

T

TensorFlow software library (Google), 197
testing
  interfaces to reveal metaphors for interaction, 25–28
  prototyping and, 179–183
  visual elements for correct identification, 16
Thinking, Fast and Slow (Kahneman), 11, 56
Tobii eye tracking technology, 13
too much information, 56–57
Torch (Facebook), 197
touchscreen tests, 25–28
Tunisian ants in the desert, 19–21
Turing, Alan, 196–197
Turing test, 196
Tversky, Amos, 47

U

unconscious behaviors, 12–14, 64, 70
usability test findings, 69–70
user research
  analyzing. See Post-It note categorization method
  appeal form of emotion and, 130–131, 161–163
  awaken form of emotion and, 130–131, 165–166
  case study, builders, 167–169
  case study, high-net-worth individuals, 169–170
  case study, let's just build it!, 178–179
  choosing contextual interviews, 63–65
  common questions, 74–75
  empathy research, 65–70
  enhance form of emotion and, 130–131, 163–165
  getting to deep desires, goals, and fears, 55, 59–60, 131–133
  recommendations for, 70–74, 171
  Six Minds considerations, 61–62

US penny, 31, 39

V

virtual reality (VR), 24
virtual space, wayfinding in, 21–29, 104–107
vision, attention, and automaticity
  about, 4, 7, 9–12, 78
  appeal form of emotion and, 162
  case study, auction website, 90–91
  case study, builders, 167
  case study, high-net-worth individuals, 169
  case study, security department, 87–88
  case study, website hierarchy, 89–90
  Double Diamond process and, 175
  empathy in design decisions, 187
  eye tracking. See eye movement and tracking
  machine learning/artificial intelligence and, 200–201
  null result, 15
  Post-It note categorization method, 75, 79–81, 91–94
  questions regarding customers, 61, 83–84, 90
  recommendations for, 70, 94
  See/Feel/Say/Do chart and, 158
  testing visual elements, 16
  unconscious behaviors, 12–14
  visual acuity, 16–17
  visual popout, 14–15
voice-activated interfaces, 28–29, 198–200
Voltaire (writer), 41

W

wayfinding
  about, 4, 7, 78
  audience segmentation and, 149–151
  case study, distracted movie watching, 109–110
  case study, search terms, 105
  case study, shopping mall, 104–105

[About the Author]

John Whalen has a PhD in cognitive science and over 15 years of human-centered design experience. As the lead of psychological insights and innovation at Brilliant Experience, he uses his unique blend of psychology, design thinking, and lean startup techniques to uncover business opportunities and design solutions for Fortune 100, nonprofit, and startup clients. John is a frequent presenter at conferences and a past president of the User Experience Professionals Association in Washington, D.C. His current practice focus is cognitive design—the art and science of harnessing cognitive psychology to understand users, inform design, and create compelling products and services.

[Colophon]

The animal on the cover of *Design for How People Think* is a Victoria crowned pigeon (*Goura victoria*). The species gets its name from Queen Victoria of Britain, who oversaw her empire's annexation of its habitat in Papua New Guinea. These ground-dwelling birds live in the island's marshes and forests. Habitat destruction and human predation have made the Victoria crowned pigeon a near-threatened species. It is the extinct dodo bird's closest surviving relation.

A white-tipped crest of lacy feathers distinguishes Victoria crowned pigeons from other pigeons. They have blue-gray plumage, red irises, orange feet, and curved beaks. Their heads are small relative to their bodies, and their bodies are the largest of all pigeons. As adults, these birds can weigh 5 lb and span 26–29 in. Though they have working wings, Victoria crowned pigeons fly rarely. They feed on fallen berries, seeds, and insects, so they never have to leave the ground for food.

In captivity, Victoria crowned pigeons can live 30–40 years. They mate for life, attracting a partner with bows and a display of that crest of feathers. Females lay one egg at a time and must incubate for about 30 days, a responsibility they share with their mates. Unlike most birds, hatchling pigeons will feed on milk produced by their parents in their first few days.

Victoria crowned pigeons feed, perch, and roost in groups of 2–10. Their nests are made of twigs, leaves, and roots, and they begin each morning by crying out together with their booming bird call.

Many of the animals on O'Reilly's covers are endangered; all of them are important to the world. To learn more about how you can help, go to *animals.oreilly.com*.

The cover illustration is by Karen Montgomery, based on a black and white engraving from *Brehms Thierleben*. The cover fonts are Gilroy Semibold and Guardian Sans. The text font is Adobe Minion Pro; the heading font is Adobe Myriad Condensed; and the code font is Dalton Maag's Ubuntu Mono.

Milton Keynes UK
Ingram Content Group UK Ltd.
UKHW052251270924
448904UK00009B/88